Current Topics in Pathology

Continuation of Ergebnisse der Pathologie

62

Editors

E. Grundmann · W. H. Kirsten

Developmental Biology and Pathology

Contributors

C. R. Austin, H. Beier, O. Bomsel-Helmreich, A. Boué, J. G. Boué, H. W. Denker, W. Engel, Ch. J. Epstein, V. H. Ferm, W. Franke, A. Gropp, M. Karkinen-Jääskeläinen, C. Lutwak-Mann, B. Putz, I. Saxén, L. Saxén, H. Spielmann, D. Szöllösi, U. Zimmermann

Editors

A. Gropp · K. Benirschke

With 86 Figures

Springer-Verlag Berlin · Heidelberg · New York 1976

E. Grundmann, Professor Dr., Pathologisches Institut der Universität, Westring 17, D-4400 Münster/Westf., Germany

W.H. Kirsten, Professor Dr., Department of Pathology, The University of Chicago, 950 East 59th Street, Chicago, IL 60637, USA

A. Gropp, Professor Dr., Abteilung für Pathologie, Medizinische Hochschule Lübeck, Ratzeburger Allee 160, D-2400 Lübeck, Germany

K. Benirschke, Professor Dr., Department of Obstetrics and Gynecology, San Diego School of Medicine, La Jolla, CA 92308, USA

ISBN-13: 978-3-642-66460-1 ISBN-13: 978-3-642-66458-8
DOI:10.1007/978-3-642-66458-8

Softcover reprint of the hardcover 1st edition 1976

Preface

The early development of the mammalian embryo belongs to a period which, for the student, provides the particularly deep fascination connected with the processes of germination of the first tender buds of life. Moreover, developmental biology encompasses a very large part of biology; if broadly defined – almost all of it. The same is true for the field of pathology if the manifold possibilities of disorders of the orderly arranged pathways of developmental processes are considered. Normal development in its earliest steps – and it would be difficult to see it otherwise – means the functioning of very intricate systems of complex interdependent cycles controlled by structural, genetic, physiological and biochemical determinants. However, disturbances interfering with them in their very different ways, can lead to fetal death, disorders of growth and differentiation, malformation and disease, sometimes as late as in the next generation or later. This is, indeed, the concern of the pathologist to whom and to whose interest in developmental pathology, this book is dedicated.

The outlines of the present volume were conceived at a symposium on "Control of early embryogenesis and factors responsible for failure of embryonic development" held May 1–4, 1974 in Travemünde and sponsored by the Deutsche Forschungsgemeinschaft. Almost fifty active participants attended this conference. At the time and in keeping with the purpose of the conference, publication of the proceedings was not envisaged. However, shortly later, the recognition of the impact of what had been reported and discussed caused a growing feeling that a summing up of some of the topics treated at the conference was necessary. In selecting the material to be included in this book we were governed by the particular desire to meet some urgent fundamental needs of developmental pathology and to strengthen the experimental and theoretical basis in order to be better prepared for the practical requirements in this field.

We would like to thank the authors for their contributions to this volume. Many thanks are due to Dr. *Heinz Götze* and to Prof. Dr. *E. Grundmann* for supporting and accepting the idea to select problems of developmental biology and pathology for emphasis in this series of Current Topics in Pathology.

Lübeck — A. Gropp

San Diego — K. Benirschke

Contents

List of Contributors

C.R. Austin, Physiological Laboratory, Downing Street, Cambridge, CB2 3EG, England

H. Beier, Abteilung Anatomie der Med. Fakultät der Rheinisch-Westfälischen Technischen Hochschule, Melatener Straße 211, D-5100 Aachen, Germany

O. Bomsel-Helmreich, Laboratoire du service de gynécologie obstétrique, Hôpital Antoine Béclère, 157, Rue de la Porte de trivaux, 92140 Clamart, France

A. Boué, Centres d'Etudes de Biologie Prénatale, I.N.S.E.R.M. Groupe U.73, Château de Longchamp, 75016 Paris, France

J.G. Boué, Centres d'Etudes de Biologie Prénatale, I.N.S.E.R.M. Groupe U.73, Château de Longchamp, 75016 Paris, France

H.W. Denker, Abteilung Anatomie der Rheinisch-Westfälischen Technischen Hochschule, Melatener Straße 211, D-5100 Aachen, Germany

W. Engel, Institut für Humangenetik und Anthropologie der Universität, Albertstraße 11, D-7800 Freiburg, Germany

Ch.J. Epstein, Departments of Pediatrics and of Biochemistry and Biophysics, University of California, San Francisco, CA 94143, USA

V.H. Ferm, Department of Anatomy/Cytology, Dartmouth Medical School, Hanover, NLL 03755, USA

W. Franke, Institut für Humangenetik und Anthropologie der Universität, Albertstraße 11, D-7800 Freiburg, Germany

A. Gropp, Abteilung für Pathologie der Med. Hochschule Lübeck, Ratzeburger Allee 160, D-2400 Lübeck, Germany

M. Karkinen-Jääskeläinen, III Department of Pathology, University of Helsinki, 00290 Helsinki 29, Finnland

C. Lutwak-Mann, Unit of Reproductive Physiology and Biochemistry, Downing Street, Cambridge, England

B. Putz, Abteilung für Pathologie der Med. Hochschule Lübeck, Ratzeburger Allee 160, D-2400 Lübeck 1, Germany

I. Saxén, III Department of Pathology, University of Helsinki, 00290 Helsinki 29, Finnland

L. Saxén, III Department of Pathology, University of Helsinki, 00290 Helsinki 29, Finnland

H. Spielmann, Pharmakologisches Institut, Abteilung Embryonal-Pharmakologie, Thielallee 69-73, D-1000 Berlin 33, Germany

D. Szöllösi, Laboratoire de Physiologie des Poissons, I.N.R.A., 78350 Jouy-en-Josas, France

U. Zimmermann, Abteilung für Pathologie der Med. Hochschule Lübeck, Ratzeburger Allee 160, D-2400 Lübeck, Germany

I. Introduction

Developmental Anomalies Arising from Errors of Fertilization and Cleavage

C.R. AUSTIN

The process of fertilization is a complex series of events, especially when the term is used to include the maturation of the oocyte, and there are numerous points at which irregularities can occur. Errors of fertilization can also arise from anomalies in spermatozoa (and *Cohen*, 1973, adduces evidence that indeed most spermatozoa are abnormal), and so sperm maturation contributes a further quota of intricacy to the picture. Additional possibilities arise during cleavage. The well-recognized consequence is a variety of chromosomal errors – a hazard that has been appreciated at least since the work of the *Hertwigs* on the sea urchin nearly a century ago.

Many human spontaneous abortuses have clear-cut chromosomal errors (about one-third, according to *Carr*, 1972, but Drs. *Boué* and *Boué*, 1976, find an even higher proportion). On a population basis this death rate of embryos and fetuses is enormous – for example, it is well over one million a year in the European Economic Community alone. The cause of death of the remaining abortuses is largely unknown, but probably includes lethal gene mutations and environmental agents such as climatic extremes, nutritional factors, and drugs (either directly on the embryo or through action on the maternal system), and antigenic incompatability between mother and fetus. The most commonly identified chromosomal anomalies are the aneuploidies, but translocations, deletions and duplications, as well as mosaicism and chimerism, have also been reported. These defects cause the deaths of human embryos and fetuses early in pregnancy – most occur in the first trimester and certainly between the 3rd and 20th weeks of pregnancy. On the experimental side, the most impressive data are those gained by Drs. *Ford* and *Gropp* in their investigations on the *Mus musculus* x *M. poschiavinus* hybrids (details of which are discussed elsewhere in this volume). The vulnerable period includes that of organogenesis, when important embryogenic processes are underway, such as the establishment and growth of clones and the occurrence of morphogenetic cell movements. From these observations it is becoming much more evident why certain anomalies arise at certain times, and yet it is a little surprising that the death of the embryo or fetus should so frequently ensue. A very wide range of chromosomal states can be tolerated by cells in tissue culture, and many teratogenic agents are known that disturb normal development but do not kill the embryo. Perhaps the lethal outcome is because the organism cannot function adequately as a whole – and yet decapitated fetuses can survive to term (*Jost*, 1948). However that may be, we have no grounds for complaint here as far as the human subject is concerned, since it is better that the embryo should cease to exist than persist to become a congenital defective.

Another point of contrast is seen in the difference between the high sensitivity of the postimplantation embryo and the very resistant nature of the preimplantation embryo, the development of which may be interrupted by certain mutant genes (see *McLaren*, 1976), but is quite undisturbed by chromosomal error. The work of Dr. *Ford* and his colleagues in recent years strongly supports the idea that the viability and functional capacity of gametes as well as preimplantation embryos are remarkably independent of even gross genome unbalance (recently reviewed by *Ford*, 1975). These observations provide good reasons for doubting claims that selection against unbalanced genomes can occur in these stages, despite the apparent validity of supporting data. Possibly support can be found in a few instances where morphologic anomaly happens to be associated (not necessarily causally), for *Krzanowska* (1974) reports clear indications that mouse spermatozoa with seriously mishapen heads are preferentially excluded from the oviduct.

That normal bipolar spindles should be formed regardless of whether the embryo is haploid, diploid, triploid, or tetraploid, and of whether it is the maternal or the paternal chromosomal complement that is duplicated, is extraordinary. By contrast, if sea urchin eggs undergo polyspermic fertilization they form multipolar spindles (as shown as long ago as 1887 by the *Hertwigs*), and polyspermic frog eggs produce multiple spindles (*Herlant*, 1911). The result in the sea urchin is chaotic cleavage and early death of the embryo. In the frog the first cleavage is into three or more blastomeres, which in addition may be binucleate; development is abnormal though it may sometimes go as far as the tadpole stage. The inference is that unlike sea urchin and frog eggs, the mammalian egg is not troubled by trivialities like centrioles, and indeed Dr. *Szöllösi* (1976) has shown very clearly that, at least in certain animals, nothing resembling the classical centriole can be found associated with the polar or first-cleavage spindles. But this still leaves a mystery, because whatever it is that organizes these spindles is apparently independent of ploidy, and as yet an adequate explanation as to how that can be is lacking.

The stability of the mammalian cleavage embryo testified to its remarkable regulative capacity, evidence of which is seen also in other features. Thus not only do the single blastomeres of 2-, 4- and 8-cell eggs show evidence of totipotency, but fused embryos also regulate to form a single organized structure. Even more striking, I think, is the fact that no teratogenic agent, to my knowledge, has ever been conclusively shown to affect the cleavage embryo in such a way as to lead to the ultimate appearance of a congenital defect. Either the embryo dies because too many of its cells have been injured or destroyed, or else it survives and undergoes the necessary adjustment to give rise to an apparently normal individual at birth.

A problem that also deserves attention is the nature of the cellular conditions that may be causally related to these chromosomal mishaps, and in this connection we seem to have evidence for changes associated with aging of individuals as of cells. There is, for example, the lower chiasma frequency in oocytes destined to be ovulated later in reproductive life; this change tends to cause failure of the normal meiotic disjunction of chromosomes and so is considered to lead to mono- and trisomies. If the *Henderson-Edwards* (1968) theory is correct, the effect is not to be attributed to aging of the oocyte itself but, firstly, to the tendency for there to be fewer chiasmata in oocytes that enter meiosis late and, secondly, to the ovulation of the oocytes in the actual sequence in which they entered meiosis. Consequences of cell aging appear to include some kind of deterioration in spindle function which results in failure of polar body formation and the establishment of the state of triploidy via polygyny (digyny). My own early observations on polar body formation suggested that the maturation spindle itself induces the cleavage furrow that eventually separates the polar body from the oocyte (*Austin*, 1951). An essential part of the idea was that the meiotic spindle on reaching telophase was still oriented parallel to the tangent to the oocyte surface, and then underwent rotation finishing up with a radial orientation. The cleavage furrow followed in the wake of the midbody of the spindle. Other workers, however, have maintained that the spindle, at least in some species, already lies in a radial orientation when it reaches telophase and shows no sign of rotation; if this were true it would be more difficult to explain the involvement of the spindle in polar-body formation. By the same token this alternative theory robs us of a possible explanation for polar body failure with oocyte aging – if spindle rotation is normal, one can propose that in the aging oocyte the polar body is not formed because the spindle fails to rotate or rotates while still in metaphase. Perhaps those who have not seen evidence of telophase rotation were observing aged eggs. There is certainly room here for further enquiry.

Another sign of deterioration in the aging oocyte is that it seems to lose one or both of the mechanisms underlying its defenses against polyspermy and the resulting triploidy. Spontaneous breakdown of cortical vesicles ("cortical granules") has been observed in hamster aging eggs, and movement away from the surface in rat and mouse aging eggs, and these things may underlie the loss of the zona reaction, but we still have no clear idea of the mechanism of the block to polyspermy nor how it changes with aging. Rather less obviously associated with aging, but certainly connected in some way with experimentally induced deterioration of the oocyte, are two further phenomena. One is the event known as "immediate cleavage," which, if followed by fertilization, can conceivably be the way in which certain chimeras could arise – several cases of human chimerism that have been reported in the medical literature may have originated in this way (see *Austin*, 1969). Immediate cleavage is evidently due to the migration of the second maturation spindle toward the center of the oocyte from which point it brings about bisection of the oocyte into two approximately equal halves (*Braden* and *Austin*, 1954). Such a process was perhaps first described by *Conklin* (1917) who centrifuged *Crepidula* eggs so as to bring the meiotic spindle to the center. Some years later *Tyler* (1932) placed unfertilized *Urechis* eggs in hypotonic sea water and they underwent cleavage into two equal cells instead of emitting polar bodies – some went on to develop into embryos. But why the spindle should migrate (spontaneously or after experimental treatment) – and indeed what holds it in the cortical location under normal circumstances – are still mysteries. One may perhaps speculate that spindle movement of this kind is effected through a microtubular system similar to that which forms the sperm aster in the eggs of marine creatures and is evidently responsible for moving the male pronucleus around. Microtubules abound in the eggs of some invertebrate animals, and perhaps there are sufficient in mammalian eggs to provide an explanation of spindle movement (*Szöllösi*, 1976).

The second phenomenon connected with induced deterioration of the oocyte is the state of tetraploidy following treatment of the fertilized mouse egg just before the first cleavage division either with colchicine which inhibits spindle formation (*Pincus* and *Waddington*, 1939) or with cytochalasin B which inhibits cytokinesis but not mitosis (*Snow*, 1973). But which of these represents the course of events in spontaneous tetraploidy is not known. In passing it is of interest that several of Snow's tetraploids survived to late pregnancy, evidently immune to the high lethality of this condition in man.

Without doubt many intriguing problems await solution in this field of enquiry, solution that should surely be hastened by free informed discussion. If we do not find solutions – and that would be a great deal to expect – then let us find good reasons for their lack.

References

Austin, C.R.: The formation, growth and conjugation of the pronuclei in the rat egg. J. roy. micr. Soc. 71, 295 (1951).

Austin, C.R.: Variations and Anomalies in Fertilization. In: Fertilization, Vol. II (*C.N. Metz, A. Monroy*, eds.), p. 437. New York–London: Academic Press 1969.

Boué, A., Boué, J.: Chromosomal Anomalies in Early Spontaneous Abortion. Current Topics in Pathology (*E. Grundmann, W.H. Kirsten*, eds.), 62, 193-208 (1976).

Braden, A.W.H., Austin, C.R.: Reaction of unfertilized mouse eggs to some experimental stimuli. Exp. Cell Res. 7, 277 (1954).

Carr, D.H.: Chromosomal anomalies in human foetuses. Res. Reprod. 4, 3 (1972).

Cohen, J.: The comparative physiology of gamete populations. Advanc. comp. Physiol. Biochem. 4, 267 (1971).

Conklin, E.G.: Effects of centrifuged force on the structure and development of the eggs of Crepidula. J. exp. Zool. 22, 311 (1917).

Ford, C.E.: The time in development at which gross genome unbalance is expressed. In: The Early Development of Mammals (*M. Balls, A.E. Wild*, eds.). Cambridge University Press 1975.

Henderson, S.A., Edwards, R.G.: Chiasma frequency and maternal age in mammals. Nature (Lond.) 218, 22 (1968).

Herlant, M.: Recherches sur les oeufs di- et trispermiques de grenouille. Arch. Biol. (Liège) 26, 173 (1911).

Hertwig, O., Hertwig, R.: Über den Befruchtungs- und Teilungsvorgang des tierischen Eies unter dem Einfluß äußerer Agentien. Jena. Z. Med. Naturw. 20, 120 (1887).

Jost, A.: Influence de la decapitation sur le developpement du tractus genital et des surrenales de lapin. C.R. Soc. Biol. (Paris) 142, 273 (1948).

Krzanowska, H.: The passage of abnormal spermatozoa through the uterotubal junction of the mouse. J. Reprod. Fertil. 38, 81 (1974).

Pincus, G., Waddington, C.G.: The effects of mitosis-inhibiting treatments on normally fertilized pre-cleavage rabbit egg. J. Hered. 30, 515 (1939).

Snow, M.H.L.: Tetraploid mouse embryos produced by cytochalasin B during cleavage. Nature (Lond.) 244, 513 (1973).

Szöllösi, D.G.: Oocyte Maturation and Paternal Contribution to the Embryo in Mammals. Current Topics in Pathology (*E. Grundmann, W.H. Kirsten*, eds.), 62, 9-27 (1976).

Tyler, A.: Changes in volume and surface of Urechis eggs upon fertilization. J. exp. Zool. 63, 155 (1932).

II. Oocyte, Early Embryo and Maternal Host. Morphology and Biochemistry

Oocyte Maturation and Paternal Contribution to the Embryo in Mammals

D. SZÖLLÖSI

The progressive events taking place in the nucleus from the intact germinal vesicle stage to the second meiotic metaphase will be referred to here as oocyte maturation. This maturation, which is under the influence of gonadotrophins, is a prerequisite for rendering an oocyte fertilizable by a spermatozoon. Whether the subsequent cleavage and development is normal relies heavily on the existence of cortical granules and on the secretion of their contents at the moment of sperm attachment. The development of a defense mechanism to avoid polyspermy is initiated during the oocyte growth phase. Even though in most cases the spermatozoon is incorporated entirely during its penetration process, the question remains if in addition to the nucleus, other organelles entering with it remain physiologically active. In this regard, the mitochondria with their known mechanism for self-replication, and centrioles, which in somatic cells duplicate themselves, are of particular interest. The developmental changes of some organelles will be discussed under normal and certain pathologic conditions. The more recent findings of virus-like "A" particles will also be presented.

I. Oocyte Maturation

During the latter embryonic period the oogonia differ from somatic cells only in minutia. They divide mitotically and possess the usual cell organelles. They are of a large size measuring 30-50 μm and the nuclear-cytoplasmic ratio is great. At about birth or shortly thereafter the meiotic chromosomes with their characteristic synaptonemal complexes are recognizable. Close to the nuclear envelope at this stage, a pair of centrioles can be found. This holds true also for oocytes in their early growth phase (*Anderson, E.*, 1970). The centrioles disappear, however, during later growth and are not present during oocyte

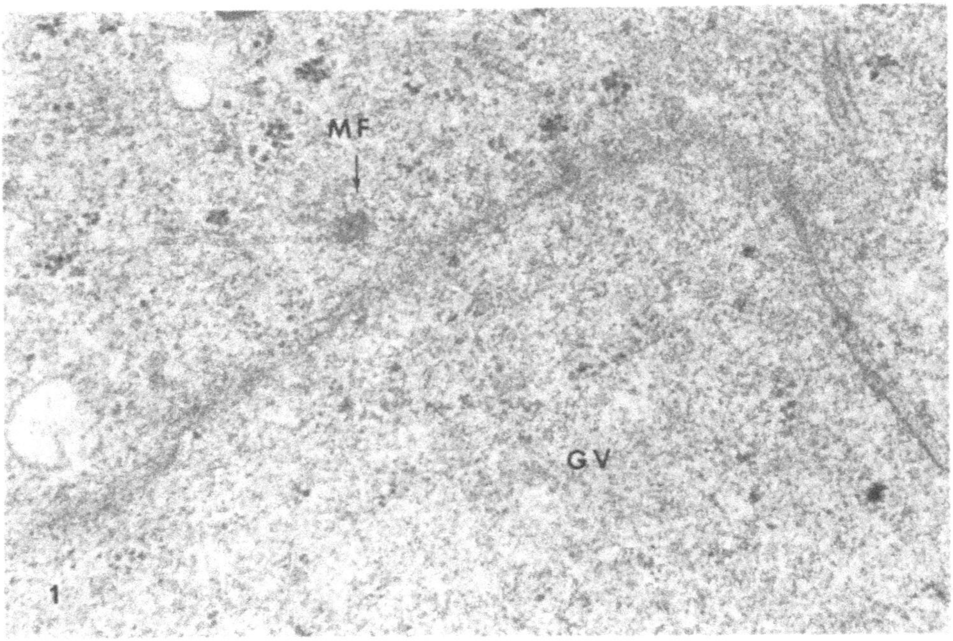

Fig. 1. In oocytes cultured for brief periods of time, small, dense, filamentous structures (MFs) appear close to germinal vesicle (GV) from which microtubules radiate. 56,000 x

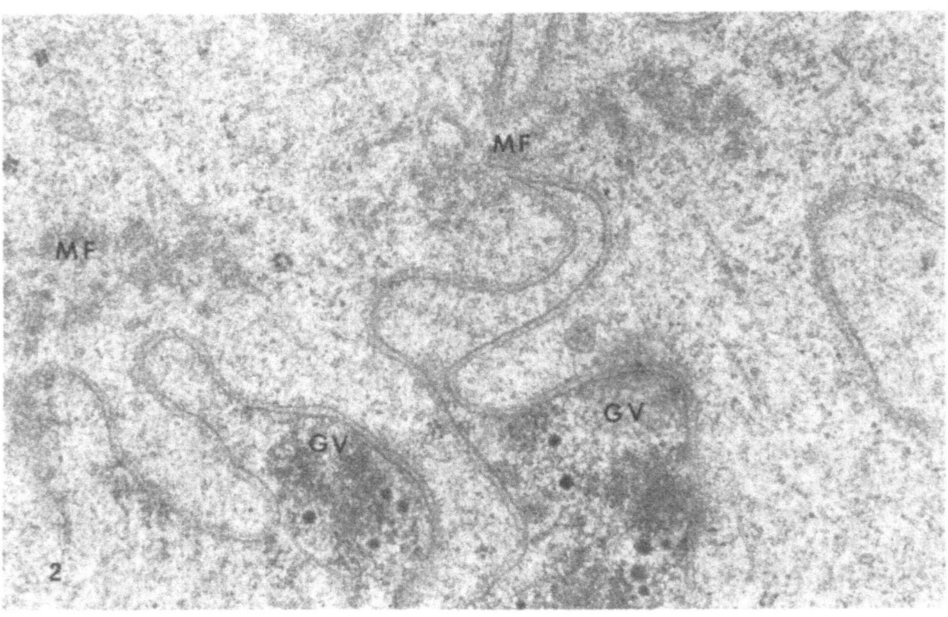

Fig. 2. *In vitro* cultured oocytes show extensive microtubule foci (MFs) near GV. From these, microtubules radiate into all directions but particularly toward indentations of nucleus. 36,600 x

maturation. The details of their decomposition are not known, but it remains a general observation that during the late oocyte growth and during the maturation divisions centrioles are absent. This has been analyzed in detail in the first and second maturation division in the mouse (*Szöllösi* et al., 1972a) and similar data have been supplied also for the rabbit and the rat (*Szöllösi*, 1973a, 1974).

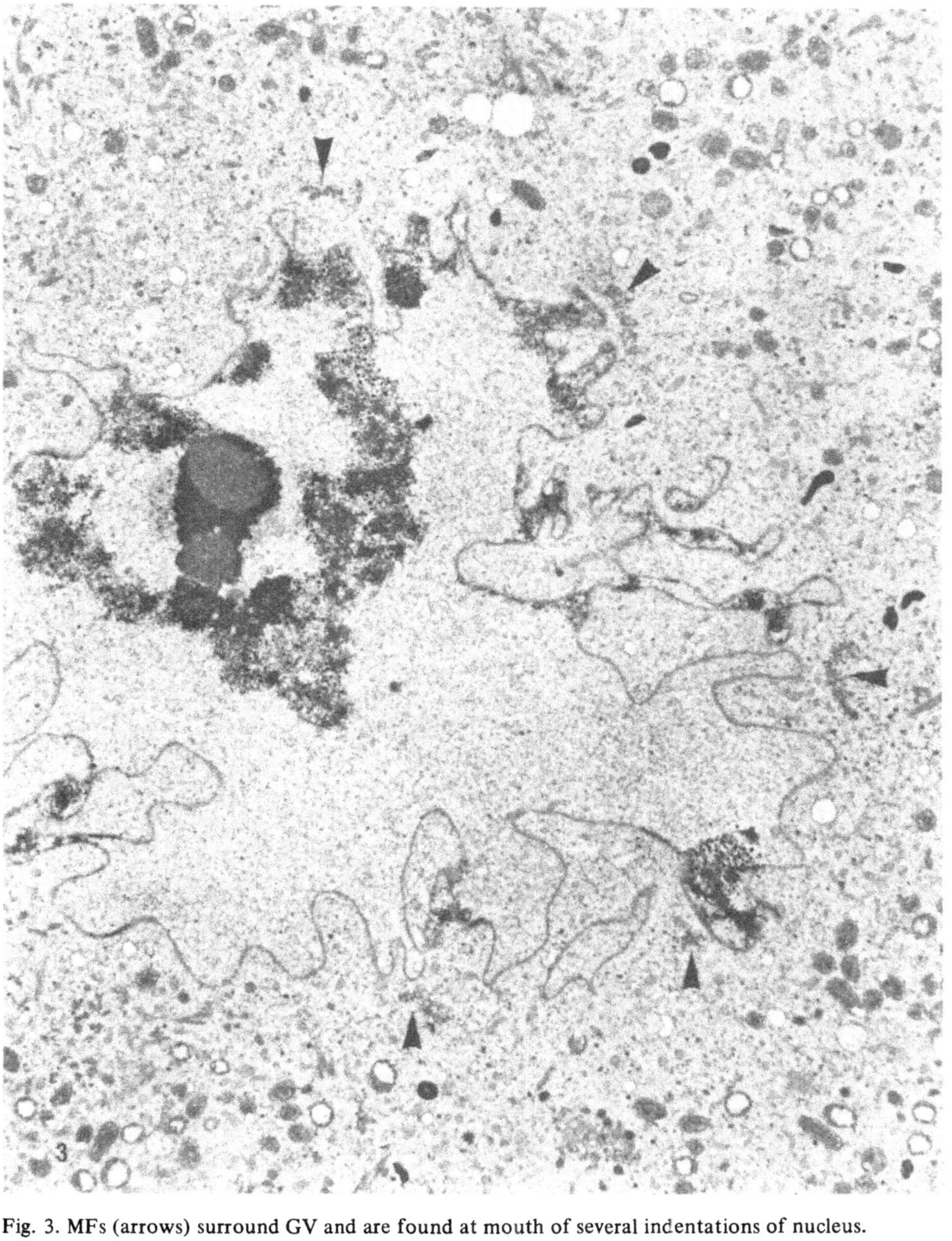

Fig. 3. MFs (arrows) surround GV and are found at mouth of several indentations of nucleus. 6,800 x

When oocytes are released from antral follicles of the mouse ovary the nucleus is nearly spherical (*Calarco* et al., 1972), and most of them undergo maturation (*Donahue*, 1968). Near the nucleus microtubules are observed in a random orientation. Small electron-dense structures appear in the proximity of the nuclear envelope that are apparently foci of microtubules (Fig. 1). When such mouse eggs are cultured under appropriate conditions (*Donahue*, 1968), the nuclear surface becomes undulative, and the microtubules oriented. Most microtubules radiate toward the nucleus (germinal vesicles, GV) from the dense foci but also toward the peripheral cytoplasm (Fig. 2). The microtubule foci (MF) or microtubule organizing centers (MTOC) (*Pickett-Heaps*, 1971) are scattered, apparently in a random manner, around the GV, and the microtubule bundles that project toward it indent it (Figs. 2, 3). The nuclear surface undulates strongly before finally some microtubules penetrate across its envelope, passing through tiny breaks. The chromosomes have contracted during the same period of time and are in close association with the nuclear envelope. Kinetochores become recognizable only after the breaks in the envelope occur (*Calarco*, 1972; *Calarco* et al., 1972). The nuclear envelope becomes more folded with time and, finally, it breaks down completely. At this time nuclear pores are completely lacking (*Szöllösi* et al., 1972b). Adjacent membrane segments of the nuclear envelope adhere to each other by the surfaces having faced the nuclear sap, forming quadruple membranes (*Calarco* et al., 1972; *Szöllösi* et al., 1972b). The MFs are still scattered around the nuclear sphere.

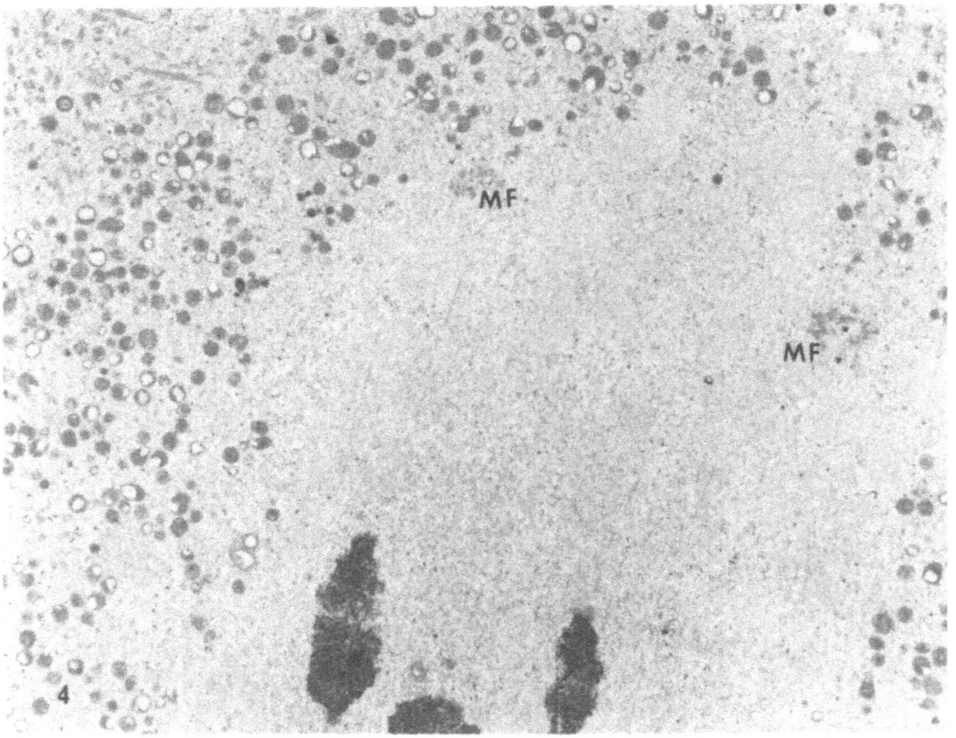

Fig. 4. Metaphase I spindle of mouse oocytes is barrel-shaped. Microtubules project to MFs. 5,900 x

The spindle polarizes subsequently and the microtubules align along an axis. There are many continuous and kinetochore fibers. These microtubules do not project toward a well-defined pole with a centriole at its center, but to several MFs and the spindle remains

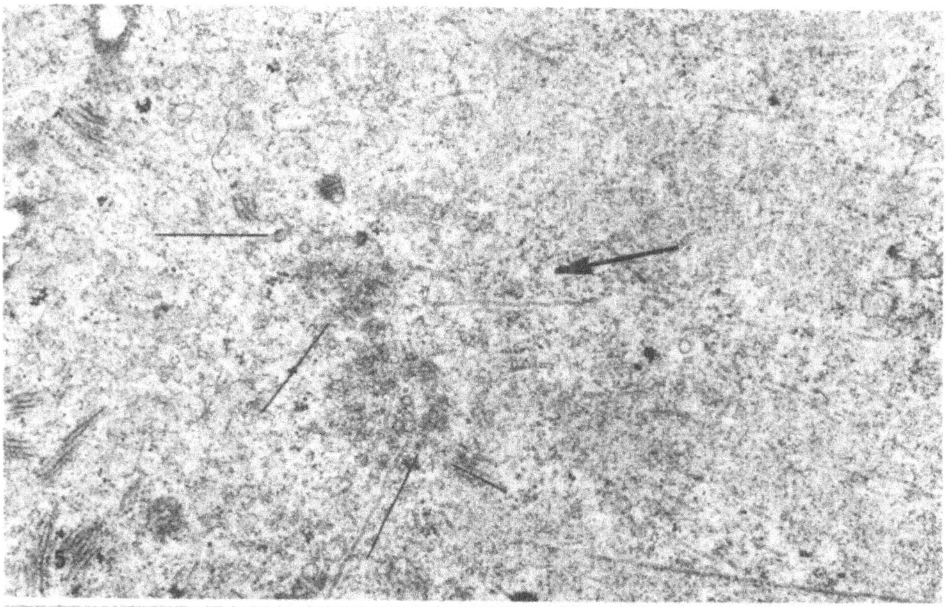

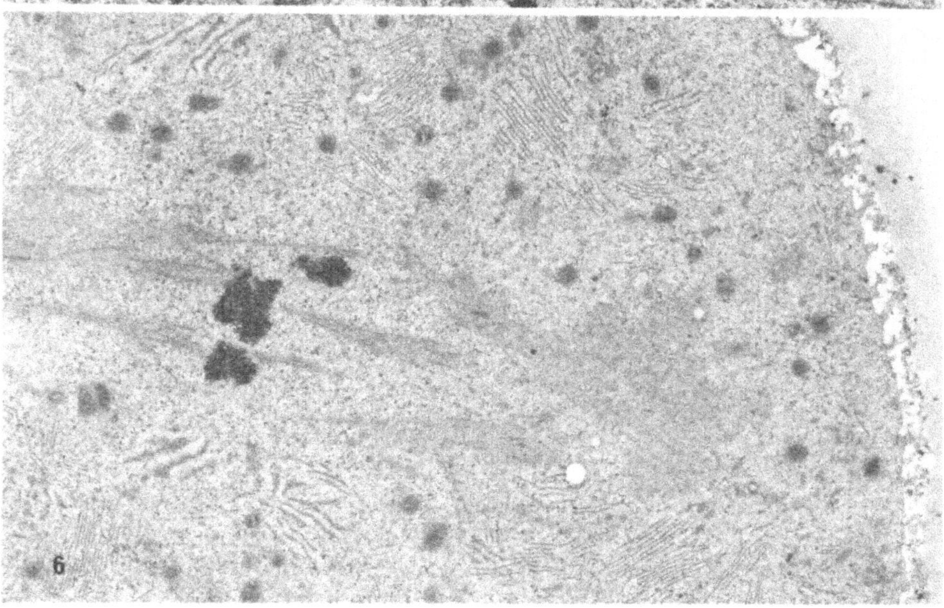

Fig. 5. MFs represent small asters. Microtubules project primarily toward body of spindle (large arrow) but also in all directions (small arrows). 22,500 x

Fig. 6. At spindle pole of second meiotic spindle of recently ovulated rat oocytes, MF is represented by large mass of filamentous material of medium electron density. 4,900 x

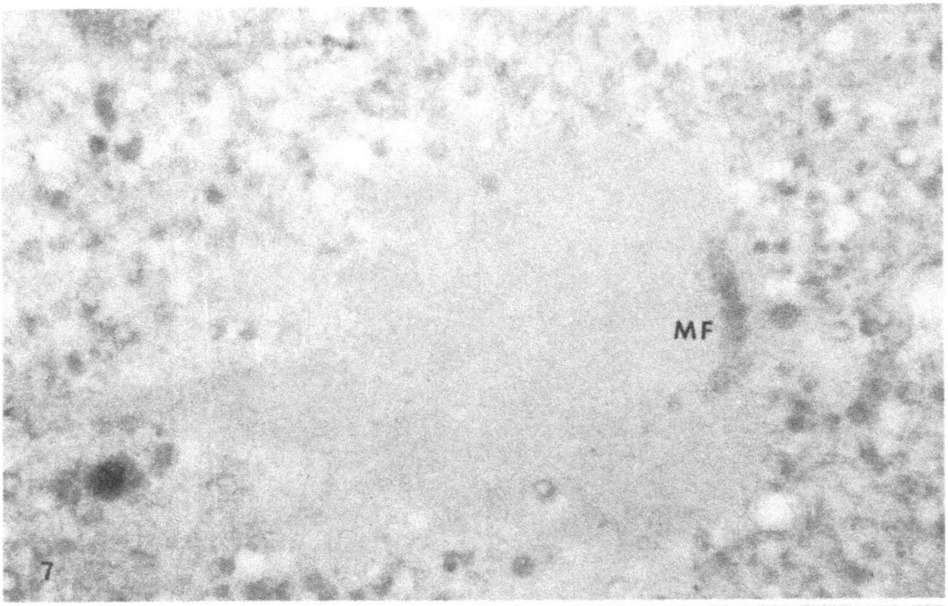

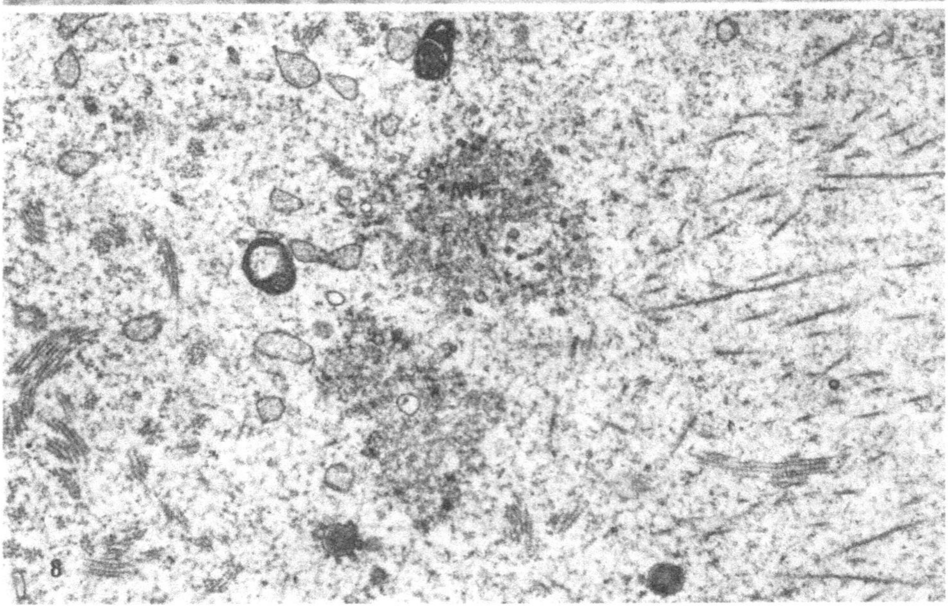

Fig. 7. At pole of 2nd meiotic spindle in one micron thick section (viewed with high voltage electron microscope) MFs are represented by fused, cap-like structure. 5,400 x

Fig. 8. Two closely placed MF complexes represent spindle pole in this thin section at first cleavage division. Many small vesicles are enclosed in electron dense filamentous masses. 6,7000 x

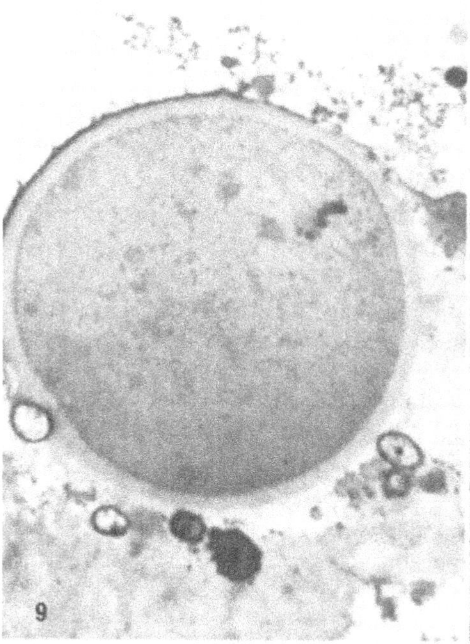

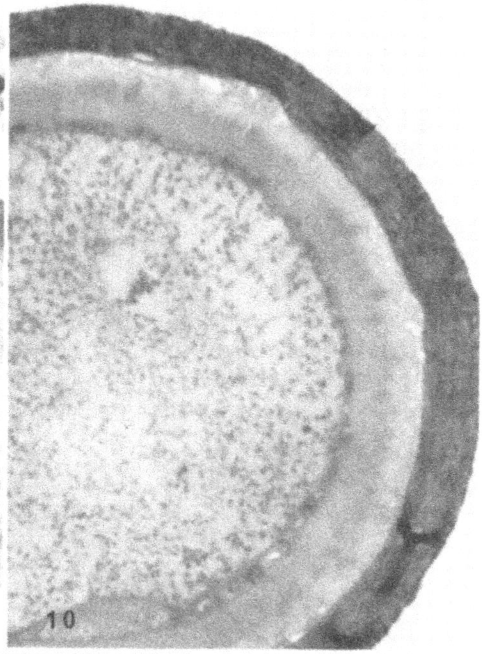

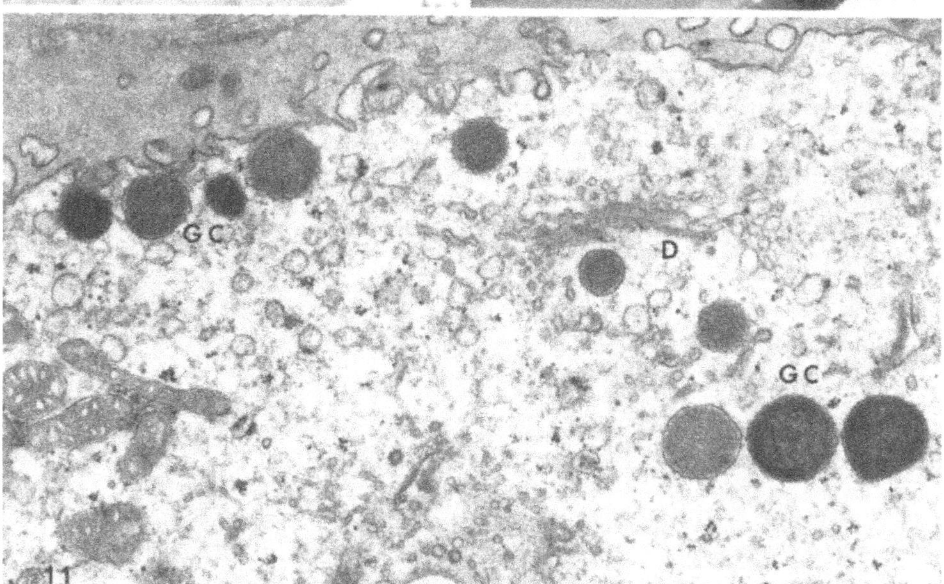

Fig. 9. Second meiotic spindle is oriented paratangentially to oocyte surface (rat). 700 x

Fig. 10. In tubally aged eggs spindle often migrates centrally in ooplasm (rabbit oocyte). 700 x

Fig. 11. Several cortical granules (CG) of different sizes lie near dictyosome (D) apparently during formative phase. 28,800 x

barrel-shaped (Fig. 4). At each pole a number of MFs are found, each appearing as small asters (Fig. 5). The largest number of microtubules project, however, toward the body of the spindle and are parallel to the overall spindle axis. Rarely, some traverse it obliquely. The number of MFs may be variable at each pole. They may be able to fuse with each other or multiply by fragmentation. Most chromosomal tubules appear to project toward one MF but it is possible that they may be anchored at two or more. Conversely, it is also feasible that spindle tubules project from several chromosomes to one single MF.

Subsequent to the separation of the chromosomes and formation of a telophase chromatin mass, the second spindle is rapidly established. Again the spindle is barrel-shaped even though the pole regions are narrower than in the first one. The MFs appear to merge into one or to a few larger masses serving as foci for most spindle microtubules (Fig. 6). From studies with the high voltage microscope of serial one micron sections, the bulk of the MFs appears to form a cap-like density at the poles (Fig. 7). Under no circumstances was a centriole found in either the first or second maturation spindle.

When these studies were extended to cleavage stages, the spindle structure of the first and second cleavag divisions (Fig. 8) were found to be similar to that of the second meiotic spindle; it is a tight, pointing barrel. No centrioles were found, as yet, at these developmental stages either. Centrioles were found, however, in the cells following the third cleavage division. It is not known if centrioles appear synchronously in each cell of the embryo or if it depends on the individual cell and its function. (See further discussion regarding this point in the section on sperm components in the egg.)

The spindle without centrioles may be less stable and, in aged eggs, it often disperses, which may result in the formation of a multiplicity of subnuclei, a feature frequently encountered in mammalian species. The second maturation spindle, which is oriented paratangentially to the egg surface at time of ovulation, often undergoes a rotation by 90° and migrates toward the center of the oocyte (Figs. 9, 10). If the spindle disintegrates, the chromosomes often form a large pronucleus-like structure, which is effectively a diploid cell (*Szöllösi*, 1971, 1973a, 1974) resulting from the suppression of polar body extrusion.

II. Cortical Granules in Mammalian Eggs

At time of sperm attachment, a set of cortical granules or vesicles is extruded in a number of invertebrate and vertebrate eggs, a reaction referred to as the cortical reaction. A role in blocking polyspermy has been ascribed to this reaction. Cortical granules exist also in mammalian eggs (*Austin*, 1956) and it seems that they are a general feature (*Szöllösi*, 1962, 1967). Cortical granules are generated in the multiple, widely scattered Golgi systems, although the site of synthesis of their constituents is not yet known (Fig. 11). They begin to appear at different phases of the oocyte growth in different mammals; in the rat and mouse for example, the first granules appear in a unilaminar, small follicle when the zona pellucida is only partially formed. But in many other mammals they make their first appearance in a multi-layered follicle.

As soon as the cortical granules begin to form, their migration toward the cell surface commences, but migration is not completed until the time of ovulation. In nearly mature oocytes, large local aggregations of cortical granules are found in the peripheral egg cytoplasm in certain mammalian species (Fig. 12). The granules will be distributed more or less as a monolayer only in the ovulated, unfertilized egg. The distribution of the granules is not uniform and the granule density may vary according to regions of the egg. They are

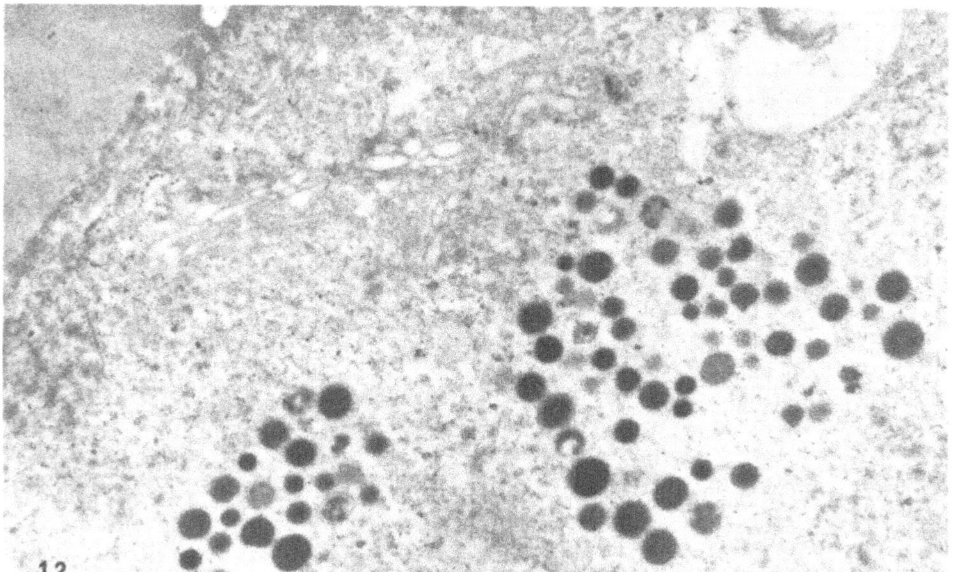

Fig. 12. Cortical granules are found in large aggregates near cell membrane in ovarian oocyte of calf. 23,400 x

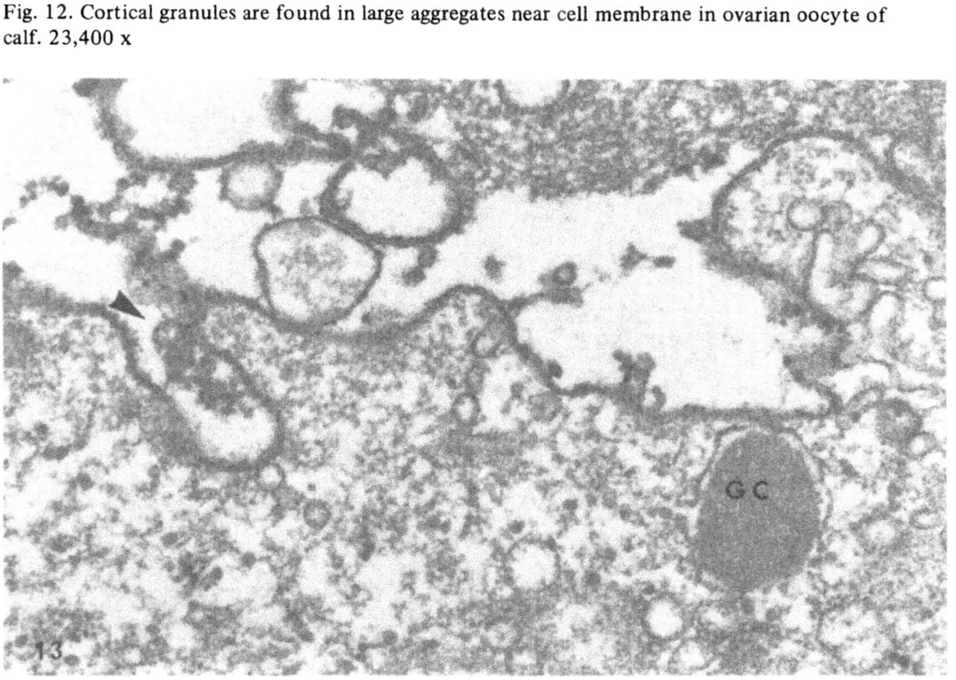

Fig. 13. Activation of egg initiates cortical reaction. In this micrograph one granule (CG) is still intact while very close by, another has extruded its contents into perivitelline space (arrow) (rat zygote). 6,100 x

usually lacking along the cell membrane overlying the second maturation spindle. The second spindle forms near the area where the first polar body was abstricted. The lack of cortical granules in that membrane region may be ascribed to currents in the cytoplasm due to cytokinetic events.

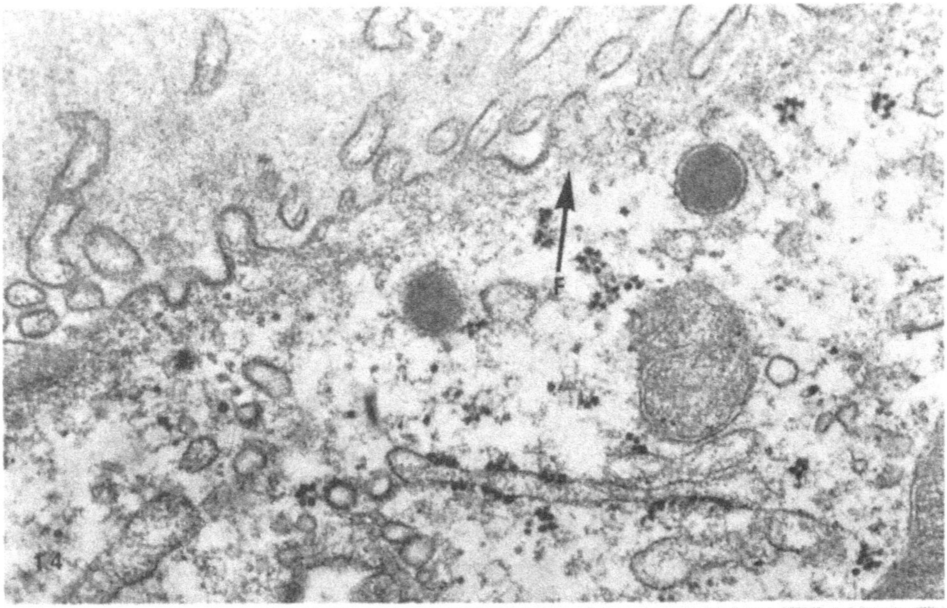

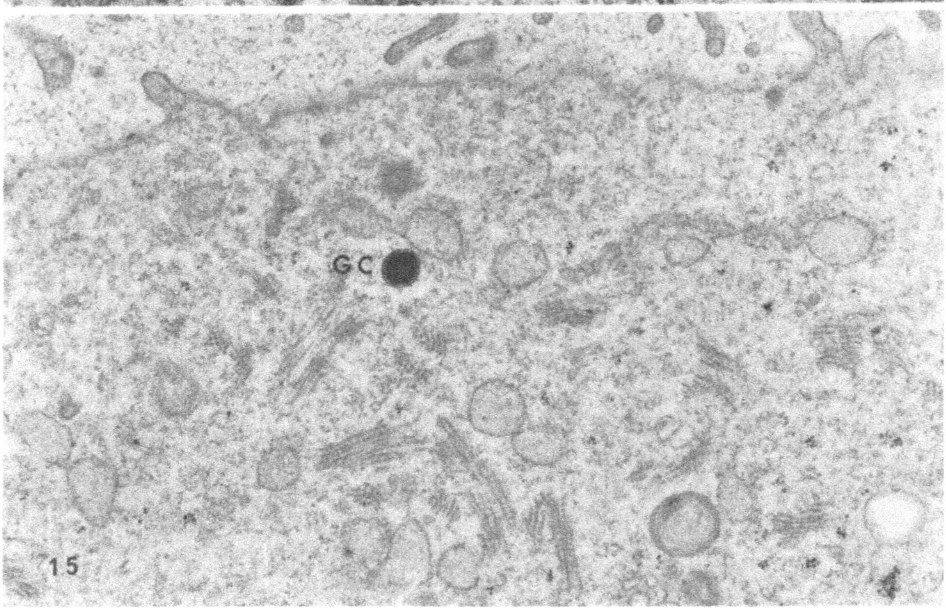

Fig. 14. Cortical filament (F) system is often observed in mammalian eggs (rat oocyte). 46,800 x

Fig. 15. Upon aging in oviduct, cortical granules of unfertilized mouse eggs leave proximity of cell membrane and migrate centripetally. 21,600 x

Little is known regarding the contents of the cortical granules. They are acid glycoproteins and react positively with the periodic acid-Schiff reaction (*Yanagimachi* and *Chang*, 1961), phosphotungstic acid stain at or below pH 1 (*Flechon*, 1970), and ruthenium red (*Szöllösi*, 1967) (Fig. 13). More recently the existence of a trypsin-like enzyme was demonstrated in them (*Gwatkin* et al., 1973).

The cortical reaction, that is the extrusion of the cortical granule contents, is induced by a specific membrane reaction: the fusion of the membrane surrounding the spermatozoon with that of the egg (*Szöllösi* and *Ris*, 1961). The reaction occurs in the manner of merocrine secretion (Fig. 14). In contrast to most glands, the secretion at the egg surface is a one-time phenomenon and not a repeatedly occurring process. The granules which are at a distance from the egg surface at this time, never react and some remain intact subsequently within the ooplasm. The actin-like microfilaments, abundant in the cortical cytoplasm, may be implicated in the expulsion and dispersion of the contents of the granules. (Fig. 15). Parthenogenetic activation by electric stimulation of hamster eggs also results in extrusion of the cortical granule contents (*Gwatkin* et al., 1973), while in rabbit eggs activated by cold shock the cortical reaction is not initiated (*Gulyas*, 1974; *Flechon* et al., 1975).

The centrifugal migration of the cortical granules is reversed in mouse eggs if they are not penetrated within a few hours after ovulation (*Szöllösi*, 1971) (Fig. 15). Under these conditions, the cortical cytoplasm of the egg becomes nearly free of the specific granules. During their centripetal migration, many of the granules swell and a paracrystalline pattern becomes visible in them. In the rat and rabbit the granules rest near the surface, but with time they appear to lose their reactivity even if a sperm would enter the egg.

Because of the time relationship between sperm attachment and the surface changes of the egg, a role in prevention of polyspermy was ascribed to the cortical reaction. It is difficult to pinpoint a time and the exact site of action. The compounds secreted into the perivitelline space may either act on reactive sites of the supplementary spermatozoa (frequent in the rabbit) on sperm receptors of the egg surface, or it may alter the external envelope of the oocyte. The portion of the zona pellucida facing the egg is changed subsequent to the entrance of the fertilizing spermatozoa in the pig for example, and supplementary sperms are arrested at that level. In the hamster it was recently found that the trypsin-like enzymes of the cortical granules remove sperm receptor sites from the zona and supplementary spermatozoa are unable to attach to it, this attachment being a prerequisite for penetration (*Gwatkin* et al., 1973). A similar enzyme system was also discovered in sea urchin eggs even if the reaction differs in detail (*Vacquier* et al., 1972; *Longo* and *Schuel*, 1973).

III. Sperm Elements in the Egg Cytoplasm

With few exceptions, it seems that the entire spermatozoon enters into mammalian oocytes (the chinese hamster and field vole can be cited with certainty as examples) (*Austin*, 1961). The sperm plasma membrane is incorporated into that of the egg with the exception of those parts that vesiculate at the time of the fusion process. Whether the sperm membrane remains later on the surface of the new embryo is not known and would be of fundamental interest. All other components of the spermatozoon enter and are immediately exposed to the ooplasm.

IV. The Mitochondrial Middle Piece

In recently penetrated eggs the mitochondria of the middle piece appear identical to those present in the mature spermatozoa (Fig. 16). The paternal mitochondria are, however,

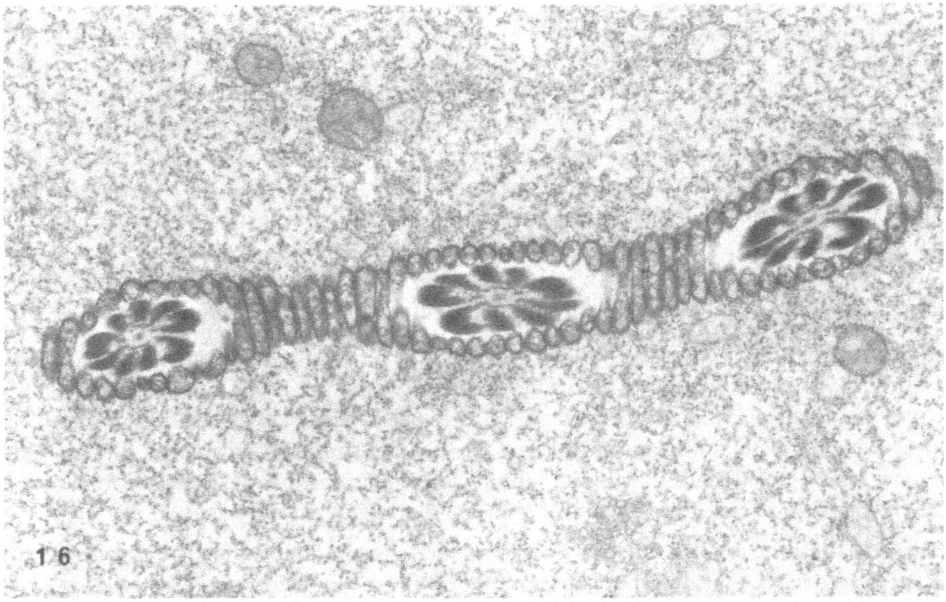

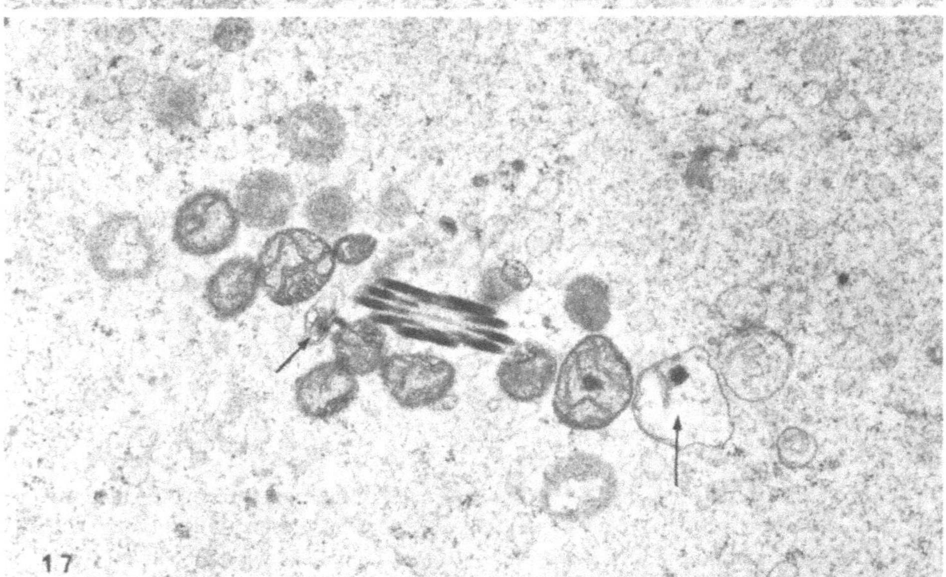

Fig. 16. Middle piece mitochondria remain unchanged shortly after sperm penetration (rat egg). 11,700 x

Fig. 17. Some mitochondria of sperm swell (arrows) already in pronucleate stage in mouse eggs, while others appear morphologically intact. 16,200 x

short-lived. Signs of swelling and degeneration appear rapidly, already during the pronucleate phase in rat (*Szöllösi*, 1967), pig (*Szöllösi* and *Hunter*, 1973), mouse (Fig. 17), and rabbit (*Szöllösi*, unpublished), but severe degenerative changes are seen during cleavage (Fig. 18).

In other zoological groups the paternally contributed mitochondria also degenerate in the embryo [sea urchin (*Anderson, W.A.*, 1968), drosophila (*Perotti*, 1973), polychete, *Armandia brevis* (*Szöllösi*, unpublished)].

In hybrid embryos produced by reverse crosses between *Xenopus muelleri* and *X. laevis* (*Dawid* and *Blackler*, 1972) mitochrondrial DNA of maternal origin could only be detected. The morphologic and biochemical studies are in agreement and indicate that sperm middle piece mitochondria do not remain physiologically functional nor is it probable that they survive. The embryo thus inherits maternally derived mitochondria only.

V. The Neck Piece and Centrioles

The studies on the lack of centrioles from the first and second cleavage spindle raised a question about the validity of Boveri's hypothesis in mammals: namely that the centrioles carried into the egg by the spermatozoon become the functional centrioles of the embryo. Already some doubts were raised about this concept when, in general, only one centriole, the proximal one, was found in mature spermatozoa (*Fawcett*, 1965). The doubts were reemphasized more recently when it was found that in mature rat sperm both centrioles were lacking (*Woolley* and *Fawcett*, 1973). In polyspermic pig eggs the proximal centriole enters the egg and can be recognized within the characteristic neck piece (Fig. 19). It is no longer recognizable, however, by the time the pronuclei swell; the space in the neck-piece at this time is void of a typical centriole, and only occasional small vesicles occupy the appropriate space (Fig. 20) (*Szöllösi*, 1973b).

VI. The Perforatorium and Tail Elements

The perforatorium or apical body is small in most mammalian species, and has not been described in fertilized eggs. In case of the rat, where the perforatorium is large, this curved, tapering, conical structure has been observed in pronucleate eggs (Fig. 21) (*Szöllösi*, unpublished). In cleavage stages it was never seen. While the exact fate of the perforatorium cannot be determined, it probably disperses in the egg cytoplasm.

The sperm tail also enters in most cases in its entirety into the egg. The external dense rods are often separated from the 9+2 microtubular elements by the time of the first cleavage or during the first few cleavages (Fig. 22). The nine dense rods remain together for varying times even if their spacial relationship is disturbed. They can be detected during cleavage stages but they have been most likely fragmented. The tail sometimes forms a bridge between adjacent blastomeres during early cleavage stages. The details of degeneration of the components of the flagellum have not been studied in sufficient detail. The microtubules retain, for a while, their 9+2 arrangement (Fig. 23), but the doublets become dispersed and later disappear.

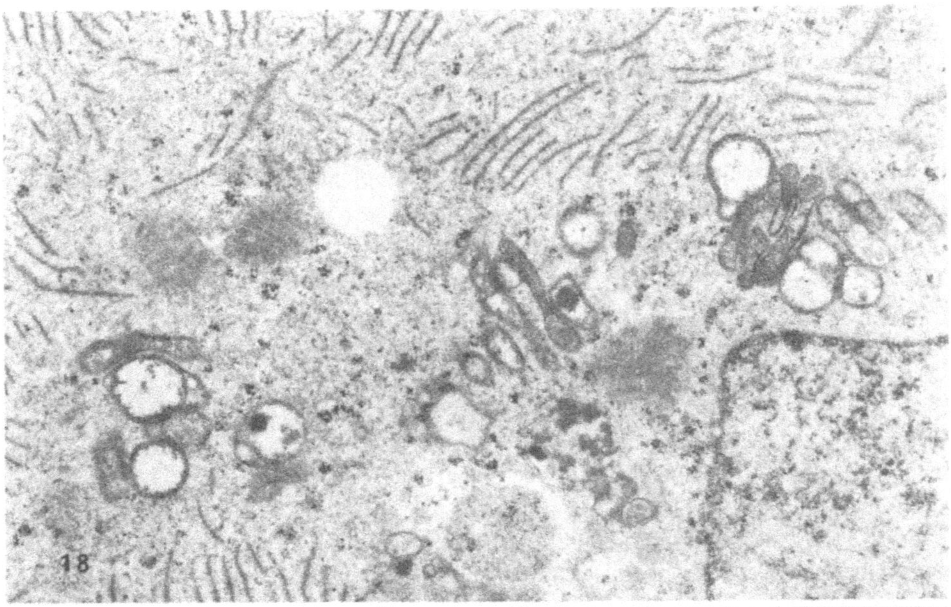

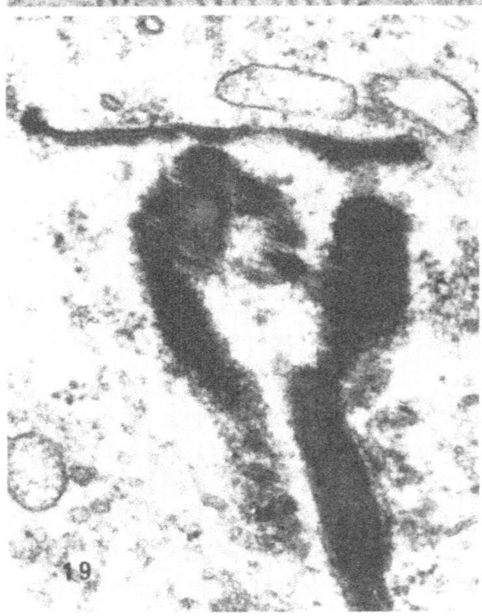

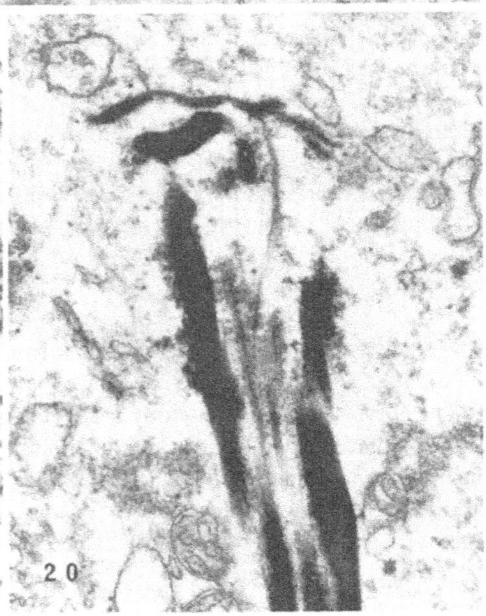

Fig. 18. Groups of swollen mitochondria remain in clusters even though they were separated from axial filament (rat two cell stage). 23,400 x

Fig. 19. In neck piece of sperm recently penetrated in pig egg, proximal centriole is recognizable. 49,500 x

Fig. 20. Proximal centriole of spermatozoon disappears from its usual location in pig eggs by time pronuclei swell and enlarge. 32,400 x

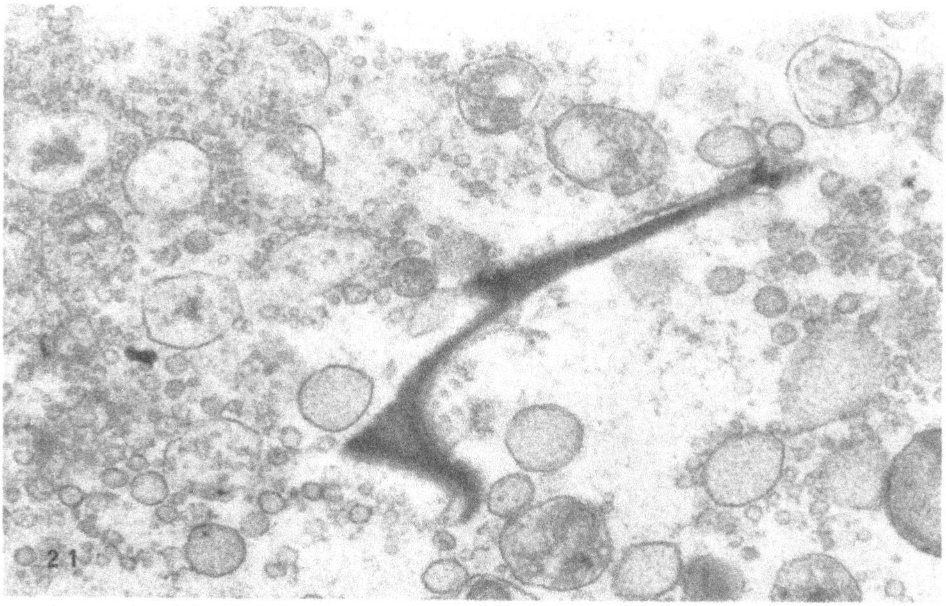

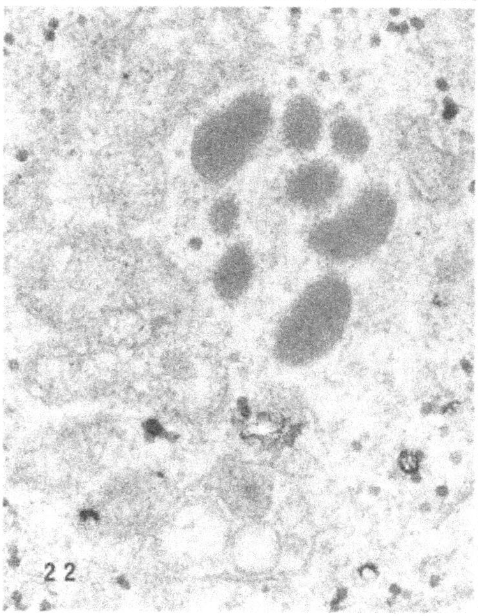

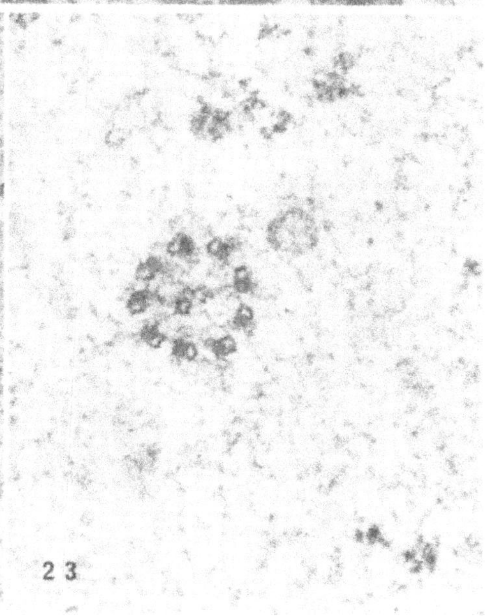

Fig. 21. Perforatorium of apical body is slightly curved, forked electron dense structure in pronucleate rat egg. It is surrounded by number of multivesiculate bodies. 24,300 x

Fig. 22. In two cell rat egg 8 of 9 dense rods remained together and are slightly displaced from their usual pattern. 9+2 microtubules are not found any longer with dense rods. Few middle piece mitochondria, some degenerating, are in proximity of rods. 56,700 x

Fig. 23. In pronucleate mouse eggs, 9+2 microtubule pattern of flagellum is retained but dense rods have already separated from them. 70,200 x

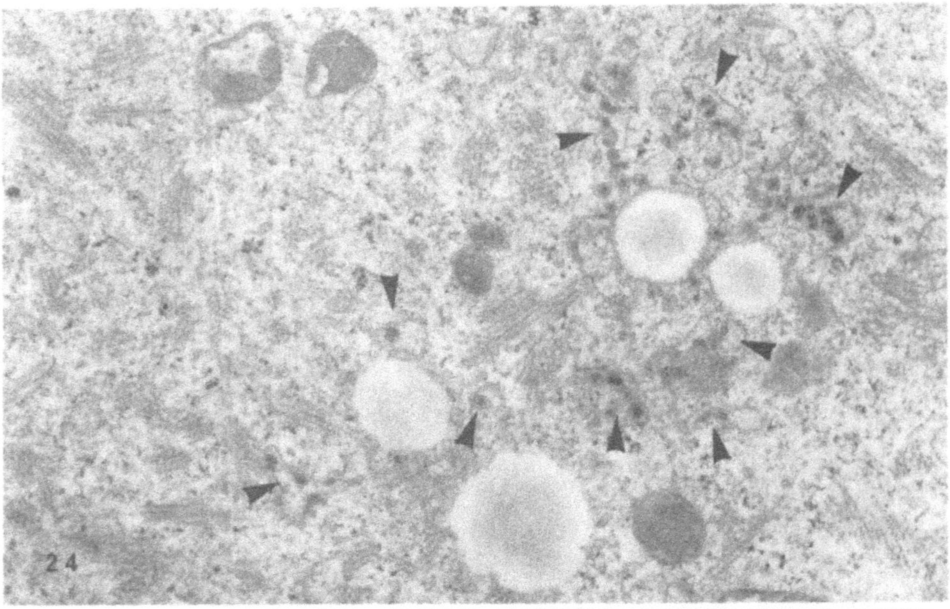

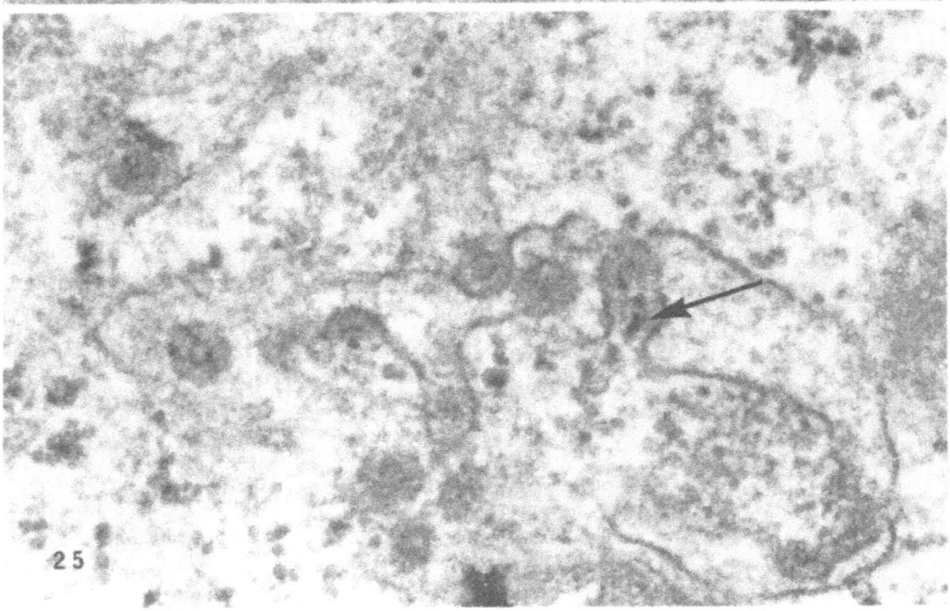

Fig. 24. Virus-like particles of A type (arrows) project into lumina of ER cisterna of NZB mouse conceptus of two cell stage. 22,500 x

Fig. 25. Several A type particles project into cistern, some forming tube-like indentations. At arrow, two nucleoids are near each other. 121,500 x

VII. Virus-like Particles

Virus-like particles, designated as A particles, by *Bernhard* (1960), are found within cytoplasmic cisternae of oocytes and in cleavage stages of mouse (Fig. 24) (*Calarco* and *Szöllösi*, 1973; *Chase* and *Piko*, 1973; *Biczysko* et al., 1973) and guinea pig (*Enders* and *Schlafke*, 1965; *Andersen* and *Jeppesen*, 1972). The particles are morphologically reminiscent of some RNA viruses found in normal and neoplastic tissues of the mouse (*Kuff* et al., 1972). The particles are either totally enclosed within the cisternae or they are budding into their lumina (Fig. 25). The buds may be very long, occasionally forming tubes, with more than one nucleoid. Even though A particles can be seen when oocytes are freshly released from the ovary, they are not present when they reach meiotic metaphase II in vivo or in vitro nor in freshly fertilized eggs. The A particles reappear in considerable numbers in the mid to late two cell stage and reach a peak at 4-8 cells. Large numbers of particles are found in every section in all strains of mice thus far studied but they are more numerous in the AKR (*Chase* and *Piko*, 1973; *Biczysko* et al., 1973) and NZB strains.

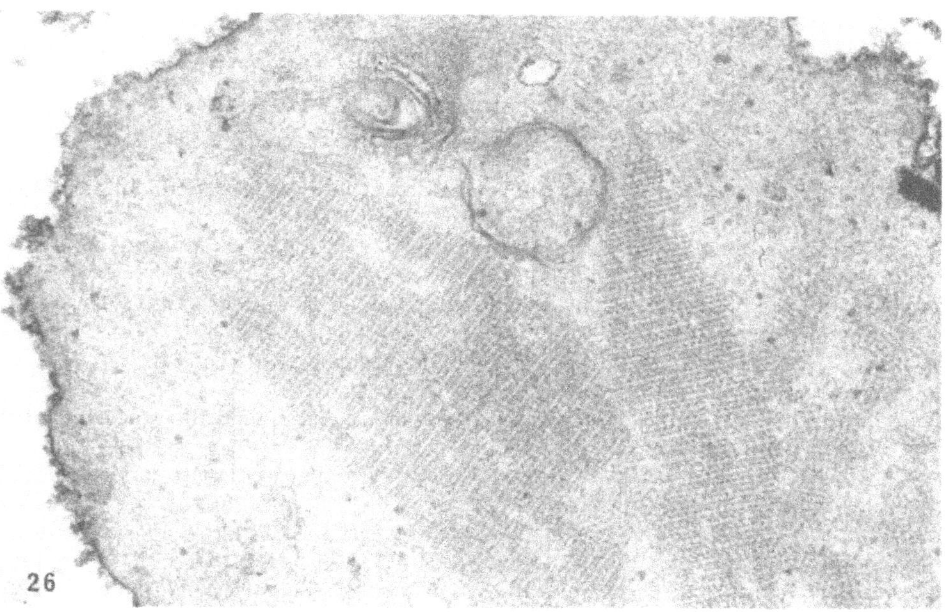

26

Fig. 26. Paracrystalline structures are frequent at two cell stage in NZB mice. 54,000 x

The frequency of the virus-like particles decreases by 16-32 cells. At two cell stages, and in the NZB strain mentioned earlier, crystalloid aggregates appear in the cytoplasm, as well as in the polar bodies (Fig. 26). The paracrystalline material is often in proximity of ER sacs. The timing of their appearance is suggestive of a relationship with the decline of the A particles but this relationship remains to be demonstrated. C-type particles were not observed in ovarian oocytes, but they were reported in blastocytes (*Chase* and *Piko*, 1973; *Biczysko* et al., 1973). Although much more work has to be done on this problem, the existence of these particles must be taken into account in all metabolic studies.

Acknowledgments: The work with the high voltage microscope has been done in collaboration with Dr. *James Laidler* at the Westinghouse Hanford Company, Richland, Washington. The present investigations have been carried out at Laboratoire de Physiologie Animale, INRA, 78350 Jouy-en-Josas and at the Laboratoire d'Histologie-Embryologie, Université de Paris-Sud, 94270 Kremlin-Bicêtre.
Certain Figures were published previously in individual articles. Many thanks are due for permission to publish: Figs. 1, 2, 3, 4 and 14 to the Cambridge University Press; Fig. 16 to the Wistar Press; Figs. 6, 7, 8, 9 and 10 to S. Karger AG, Basel.

References

Andersen, H.K., Jeppesen, T.: Virus-like particles in guinea pig oogonia and oocytes. J. nat. Cancer Inst. 49, 1403-1410 (1972).

Anderson, E.: The localization of acid phosphatase and the uptake of horseradish peroxidase in the oocyte and follicle cells of mammals (eds. *J.D. Biggers* and *A.W. Schuetz*). Oogenesis, Ch. 5, 87-117 (1970).

Anderson, W.A.: Structure and fate of the parental mitochondrion during early embryogenesis of Paracentrotus lividus. J. Ultrastruct. Res. 34, 211-321 (1968).

Austin, C.R.: Cortical granules in hamster eggs. Exp. Cell Res. 10, 533 (1956).

Austin, C.R.: The Mammalian Egg. Oxford: Blackwell 1961.

Bernhard, W.: The detection and study of tumor viruses with the electron microscope. Cancer Res. 20, 712-727 (1960).

Biczysko, W., Pienkowski, M., Solter, D., Koprowski, H.: Virus particles in early mouse embryos. J. nat. Cancer Inst. 51, 1041-1050 (1973).

Calarco, P.G.: The kinetochore in oocyte maturation (eds. *J.D. Biggers* and *A.W. Schuetz*). Oogenesis, Ch. 5, 65-89 (1972).

Calarco, P.G., Donahue, R.P., Szöllösi, D.: Germinal vesicle breakdown in the mouse oocyte. J. Cell Sci. 10, 369-385 (1972).

Calarco, P.G., Szöllösi, D.: Intracisternal A particles in ova and preimplantation stages of the the mouse. Nature (New Biol.) 243, 91-93 (1973).

Chase, D.G., Piko, L.: Expression of A- and C-type particles in early mouse embryos. J. nat. Cancer Inst. 51, 1871-1975 (1973).

Dawid, I.B., Blackler, A.W.: Maternal and cytoplasmic heritance of mitochondrial DNA in Xenopus. Develop. Biol. 29, 152-161 (1972).

Donahue, R.P.: Maturation of the mouse oocyte in vitro. I Sequence and timing of nuclear progression. J. exp. Zool. 169, 237 (1968).

Enders, A.C., Schlafke, S.J.: The fine structure of the blastocyst: some comparative studies. In: Preimplantation stages of pregnancy (eds. *G.E.W. Wolstenholme* and *M. O'Connor*). Ciba Foundation Symposium. Boston: Little, Brown 1965.

Fawcett, D.W.: The anatomy of the mammalian spermatozoon with particular reference to the guinea pig. Z. Zellforsch. 67, 279-296 (1965).

Flechon, E.: Nature glycoprotéique des granules corticaux de l'oeuf de lapine. Mise en evidence par l'utilisation comparée de techniques cytochimiques ultrastructurales. J. Microscopie 9, 221-242 (1970).

Flechon, J.E., Huneau, D., Solari, A., Thibault, C.: Réaction corticale et blocage de la polyspermie dans l'oeuf de lapine. Ann. Biol. anim. Biochim. Biophys. 15, 9-18 (1975).

Franklin, L.E., Fussell, E.N.: Evolution of the apical body in golden hamster spermatids with some reference to primates. Biol. Reprod. 7, 194-206 (1972).

Gulyas, B.J.: Cortical granules in artificially activated (Parthenogenetic) rabbit eggs. J. Anat. 140, 577-582 (1974).

Gwatkin, R.B.L., Williams, D.T., Hartmann, F.J., Kniazuk, M.: The zona reaction of hamster and mouse eggs: production in vitro by a trypsin-like protease from cortical granules. J. Reprod. Fertil. 32, 259-266 (1973).

Kuff, E.L., Lueders, K.K., Ozer, W.L., Wivel, N.A.: Some structural and antigenic properties of intracisternal A particles occurring in mouse tumors. P.N.A.S. 69, 218-222 (1972).

Longo, F.J., Schuel, H.: An ultrastructural examination of polyspermy induced by soybean trypsin inhibitor in the sea urchin Arbacia punctulata. Develp. Biol. 34, 187-199 (1973).

Perotti, M.E.: The mitochondrial derivative of the spermatozoon of *Drosophila* before and after fertilization. J. Ultrastruct. Res. 44, 181-198 (1973).

Pickett-Heaps, F.D.: The autonomy of the centriole: fact or fallacy? Cytobios 3, 205-214 (1971).

Szöllösi, D.: Cortical granules: a general feature of mammalian eggs? J. Reprod. Fertil. 4, 223 (1962).

Szöllösi, D.: The fate of sperm middle-piece mitochondria in the rat egg. J. exp. Zool. 159, 367-377 (1965).

Szöllösi, D.: Development of cortical granules and the cortical reaction in rat and hamster eggs. Anat. Rec. 159, 431-446 (1967).

Szöllösi, D.: Changes of some cell organelles during oogenesis in mammals (eds. *D.J. Biggers and A. Schuetz*). Oogenesis, Ch. 5, 47-86 (1970).

Szöllösi, D.: Cortical cytoplasmic filaments of cleaving eggs: A structural element corresponding to the contractile ring. J. Cell Biol. 44, 192-209 (1970).

Szöllösi, D.: Morphological changes in mouse eggs due to aging in the fallopian tube. J. Anat. 130, 209-225 (1971).

Szöllösi, D.: The spindle structure in mammalian eggs: the effect of aging. Proceedings of the Symposium sponsored by INSERM, DGRST, OMS (eds. *A. Boue* and *C. Thibault*) 1973a.

Szöllösi, D.: The fate of the proximal centriole of porcine sperm penetrating an egg. J. Cell Biol. 59, 342a (1973b).

Szöllösi, D.: Mammalian eggs aging in the fallopian tubes. Basel: Karger A.G. 1974.

Szöllösi, D., Calarco, P.G., Donahue, R.P.: Absence of centrioles in the first and second meiotic spindles of mouse oocytes. J. Cell Sci. 11, 521-541 (1972a).

Szöllösi, D., Calarco, P.G., Donahue, R.P.: The nuclear envelope: its breakdown and fate in mammalian oogonia and oocytes. Anat. Rec. 174, 325-340 (1972b).

Szöllösi, D., Hunter, R.H.F.: Ultrastructural aspects of fertilization in the domestic pig: sperm penetration and pronucleus formation. J. Anat. 116, 181-206 (1973).

Szöllösi, D., Hans Ris, P.D.: Observations of sperm penetration in the rat. J. Biophys. Biochem. Cytol. 10, 275-283 (1961).

Vacquier, V.D., Tegner, M.J., Epel, D.: Protease activity establishes the block against polyspermy in sea urchin eggs. Nature (Lond.) 240, 352-353 (1972).

Woolley, D.M., Fawcett, D.W.: The degeneration and disappearance of the centrioles during the development of the rat spermatozoon. Anat. Rec. 177, 289-302 (1973).

Yanagimachi, R., Chang, M.C.: Fertilizable life of golden hamster ova and their morphological changes at the time of losing fertilizability. J. exp. Zool. 148, 185-203 (1961).

Maternal Storage in the Mammalian Oocyte

W. ENGEL and W. FRANKE

I. Xenopus and Sea Urchin – An Introductory Side-glance

As early as 1896, *Wilson* stated that embryogenesis commences during oogenesis. Much evidence now exists suggesting that at least in invertebrates and lower vertebrates the information controlling early embryonic development is programmed during oogenesis (see reviews by *Davidson*, 1968; *Davidson* and *Hough*, 1972; *Gross*, 1972; *Smith*, 1972; *Gross* et al., 1973a).

Amphibian as well as echinoderm embryos may undergo early stages of development without any contributions by their genomes: inhibition of transcription by actinomycin D during cleavage does not prevent sea urchin or frog development up to blastulation (*Gross* and *Cousineau*, 1963, 1964; *Brachet* et al., 1964). Cleavage can occur in the presence of highly abnormal chromosomal complements (*Hennen*, 1963; *Briggs* et al., 1964), and even enucleated embryos develop to partial blastulae (*Briggs* et al., 1951; *Smith* and *Ecker*, 1965). Parthenogenetically activated eggs (*Terman* and *Gross*, 1966) or enucleate egg halves (*Denny* and *Tyler*, 1964) will go through early development and follow the same pattern of amino acid incorporation into protein as fertilized eggs (*Terman* and *Gross*, 1966; *Fry* and *Gross*, 1970).

Nevertheless, it is established that transcription occurs in early development in the sea urchin and amphibians (*Brown* and *Littna*, 1964; *Whiteley* et al., 1966; *Gross*, 1967a; *Tyler*, 1967; *Hynes* and *Gross*, 1972; *Gross* et al., 1973a). Among the products are mRNA molecules that enter the cytoplasm and are translated with little or no delay (*Hogan* and *Gross*, 1971). Ribosomal RNA (rRNA) and transfer RNA (tRNA) synthesis, however, begins only at the onset of gastrulation (*Brown* and *Dawid*, 1969).

1. Synthesis of Ribosome Storage

The externally developing embryo disposes of large amounts of reserve substances in the egg, which are formed during the lampbrush stage of oogenesis. In this stage, the oocyte has a large nucleus which contains 600-2000 nucleoli (*MacGregor*, 1965; *Callan,* 1966; *Perkowska* et al., 1968). These nucleoli are extremely active in RNA synthesis (*Davidson* and *Mirsky*, 1965; *Brown*, 1967; *Gall,* 1969), resulting in about 4.7 μg of ribosomal RNA in the unfertilized egg of *Xenopus laevis* compared to perhaps 40 pg in a liver cell (*Crippa* et al., 1972; *Ford*, 1972; *La Marca* et al., 1973). This goal is achieved through the selective amplification of the (28S + 18S) ribosomal genes during the pachytene stage of the meiotic prophase (*Brown* and *Dawid*, 1968; *Evans* and *Birnstiel*, 1968; *Gall*, 1968; *MacGregor*, 1968). In this way the number of ribosomal cistrons present in the oocyte nucleus increases more than a thousand times over the expected tetraploid value. The synthesized ribosomes are essentially a storage product to be used during early development (*Brown* and *Gurdon*, 1964). However, in mature oocytes and unfertilized eggs these ribosomes are inactive in protein synthesis possibly due to the presence of an inhibitory protein, acting at the monoribosome level. After fertilization this inhibitor seems to be inactivated or destroyed because of successive proteolytic degradation (*Monroy* et al., 1965; *Rinaldi* and *Monroy*, 1969; *Piatigorsky* and *Tyler*, 1970; *Metafora* et al., 1971; *Vacquier* et al., 1972; *Schuel* et al., 1973; *Gambino* et al., 1973; *Hille*, 1974; *Clegg* and *Denny*, 1974), resulting in the continuous increase in the polysome to monoribosome ratio until the mid-blastula stage (*Gurley* et al., 1973; *Balhorn* et al., 1973).

2. Synthesis of Protein Storage

The growing oocyte increases many thousand fold in volume and accumulates large quantities of protein (e.g. about 125 μg in the unfertilized egg of *X. laevis*) (*Mairy* and *Denis*, 1971; *Halberg* and *Smith*, 1975). However, about 80% of the total protein of an egg is yolk protein, which is synthesized in the liver and transported via the blood stream to the growing oocyte (*Wallace* and *Jared*, 1969; *Glass*, 1971). Protein synthesis is very low in full grown oocytes, reaches a maximum level between the I. and II. meiotic metaphase, and is constantly low or negligible thereafter. A manyfold increase in protein synthesis begins soon after fertilization (e.g., 10-15-fold in the sea urchin). Protein synthesis during oocyte maturation as well as after fertilization is due to translation of preserved messenger RNA (mRNA) rather than to transcriptional activity (*Hultin*, 1961; *Monroy* and *Vittorelli*, 1962; *Gross* and *Fry*, 1966; *Epel*, 1967; *Gross*, 1967a, 1968; *Ecker* and *Smith*, 1968; *Panje* and *Kessel*, 1968; *Crippa* and *Gross*, 1969; *Kedes* and *Gross*,1969; *Humphreys*, 1969, 1971; *Benson*, 1971; *Smith*, 1972).

3. Stable Messenger RNAs

Stored mRNA inherited from oogenesis has been demonstrated in a variety of early embryos of invertebrates and lower vertebrates (*Gross*, 1967a, 1968; *Kedes* and *Gross*, 1969). Changes in the amount, subcellular localization, and metabolism of maternal mRNAs in eggs and embryos of the sea urchin (*Slater* et al., 1973; *Wilt*, 1973; *Slater* and *Slater*, 1974;

Mescher and *Humphreys*, 1974) and *X. laevis* (*Rosbash* and *Ford*, 1974) have been observed. According to *Rosbash* and *Ford* (1974) the greatest portion of the maternally derived mRNA in the egg, characterized by a poly (A) segment at the 3'-end, is synthesized in early oogenesis during maximal formation of lampbrush chromosomes. Egg RNA/DNA hybridization experiments with embryonal RNA as competitor, revealed that a great amount of RNA sequences from the egg was still present in blastula stage embryos (*Crippa* et al., 1967; *Crippa* and *Gross*, 1969). *Davidson* and *Hough* (1971) have determined the amount of those RNA sequences in mature *X. laevis* oocytes which hybridize with unique DNA sequences. They found that at least 0.6% of the RNA from mature oocytes are complementary to 1.2% of the nonrepetitive DNA fraction of the *Xenopus* genome. If one assumes that hybridization with nonrepetitive DNA sequences is a property of mRNA, then these data can be interpreted in such a way that the complexity of the RNA in the oocyte would be sufficient to code for more than 40,000 different polypeptide chains with the size of a β-globin chain.

Thus, the mature oocyte would contain stable templates for later use, however, the diversity of functions attributable to the proteins coded for by the maternal mRNA population could hardly have been appreciated till now. Experiments with actinomycin D during early sea urchin development have given indirect evidence for the presence of maternally derived mRNA for some specific proteins. In the early embryo, microtubuline, which is a structural protein component of the mitotic apparatus, is presumably synthesized on maternally derived mRNA (*Raff* et al., 1971, 1972). This seems to be true, also, for the synthesis of histone (*Johnson* and *Hnilica*, 1971; *Crane* and *Villee*, 1971; *Ruderman* and *Gross*, 1974) and ribonucleotide reductase (*Noronha* et al., 1972). Recently, the first maternally transmitted mRNA for a definite protein has been isolated. *Gross* et al. (1973b) isolated ribonucleoprotein particles with a sedimentation constant of 20 S from the sea urchin egg. A 9 S-fraction from these particles could be translated in vitro into histone proteins. The amount of histone coding mRNA in the unfertilized egg was 25% of that present in the morula stage (*Skoultchi* and *Gross*, 1973; *Farquhar* and *McCarthy*, 1973).

4. Inactivation and Reactivation of Storage Products

A very important question is, in which way is the maternally transmitted mRNA kept inactive in the egg until later use during development. It has been proposed that the egg lacks activated tRNA (*Maggio* and *Catalano*, 1963; *Glisin* and *Glisin*, 1964), or has inactivated its ribosomal apparatus (*Hultin*, 1961; *Monroy* et al., 1965; *Piatigorsky*, 1968; *Rinaldi* and *Monroy*, 1969; *Piatigorsky* and *Tyler*, 1970). A more generally accepted model assumes that mRNA is conserved by association with proteins as ribonucleoprotein particles, so called "informosomes" (see reviews by *Spirin*, 1969, 1972; *Samarina* et al., 1973; *Gross* et al., 1973a; *Lewin*, 1974). The activation of the ribonucleoprotein particles and the onset of translational activity might then be caused by partial proteolysis during early development. It is well known that protease activity in the egg increases after fertilization (*Davidson*, 1968; *Vacquier* et al., 1972; *Schuel* et al., 1973).

II. Storage Products in the Mammalian Egg

In mammals, the nature and extent of the support furnished by the egg cytoplasm for later use in early development is largely unknown. There is no doubt that maternally derived cytoplasmic substances, formed during oogenesis, are present in the mammalian egg, but there is only indirect evidence that early development is dependent on these substances. Until now it has been impossible to demonstrate in the oocyte the presence of mRNA for a definite protein stored for later use.

Compared with the somatic cells, the mammalian egg is of enormous size, varying in diameter between about 60 μm in the mouse and 140 μm in cattle. However, compared with the egg of lower vertebrates (e.g., *X. laevis*: 850-1000 μm in diameter), the mammalian egg is so small that the amount of reserve substances in the cytoplasm should be minimal. Low concentrations of actinomycin D result in an inhibition of mouse egg cleavage (*Mintz*, 1964; *Thomson* and *Biggers*, 1966; *Golbus* et al., 1973). Therefore, it has been suggested that normal development in mammals depends upon immediate genomic activation following fertilization (see reviews by *Wolf* and *Engel*, 1972; *Church* and *Schultz*, 1974; *Epstein*, this volume). On the other hand, parthenogenetically activated mouse eggs can develop normally to the blastocyst stage (*Tarkowski* et al., 1970; *Graham*, 1970, 1974; *Graham* and *Deussen*, 1974), indicating at least that the significance of the paternal genome during preimplantation development is questionable. Furthermore, embryos with abnormal chromosomes are most probably eliminated only during or after implantation (except YO which is eliminated as early as the 2-4-cell stage; *Morris*, 1968) (*Bomsel-Helmreich*, 1971; *Ford,* 1971; *Beatty,* 1972; *Hansmann* and *Röhrborn*, 1973; *Gropp*, 1973), and also most lethal mutations in the mouse are expressed only after implantation (*Church* and *Schultz,* 1974).

1. RNA Synthesis During Mammalian Oocyte Development

Various papers have appeared wich have a bearing on RNA synthesis during oogenesis and oocyte maturation as well as on the total amount of RNA in the egg of mammals (*Deane*, 1952; *Hedberg*, 1953; *Austin* and *Braden*, 1953; *Flax*, 1953; *Odeblad* and *Magnusson*, 1954; *Reamer*, 1963; *Mintz*, 1964: *Lutwak-Mann*, 1966; *Oakberg*, 1967, 1968; *Beauregard* et al., 1968; *Baker* et al., 1969; *Manes*, 1969, 1971; *Bloom* and *Mukherjee*, 1972; *Olds* et al., 1973; *Moore* and *Lintern-Moore*, 1974; *Moore* et al., 1974; *Bachvarova*, 1974). The total amount of RNA in the mouse egg was determined by *Reamer* (1963) with 1.75 ng and recently calculated by *Olds* et al. (1973) with 0.55 ng of RNA per ovum. Quantitative estimates of RNA in the rabbit egg have been performed by *Lutwak-Mann* (1966) and *Manes* (1969). *Manes* found 20 ng of RNA per rabbit egg, which is about 35 times as much as in the mouse egg. Since the oocyte volume of the rabbit is approximately 3.5 times that of the mouse (*Brinster*, 1971a), a greater amount of RNA in the rabbit egg may be expected. However, the RNA content of the rabbit egg determined by *Manes* (1969) is still 10 times greater than that in the mouse oocyte, even if the difference in volume is taken into consideration. In this respect it should be mentioned that the small mouse embryo may not be the prototype to evaluate cytoplasmic support in early mammalian development (*Austin*, 1961; *Manes*, 1973). Instead, the mouse oocyte might represent an exception. It is not clear whether the disparity in absolute RNA content is due to a real difference between both species or due to technical difficulties. Nevertheless, the RNA content in the mammalian oocyte is very low compared to that in lower vertebrates, e.g. *X. laevis,* in which the total RNA amount has been determined with 5 μg per mature oocyte (*Ford*, 1972).

According to the photometric determinations of *Flax* (1953), the cytoplasmic content of RNA in the mouse egg augments linearly during oogenesis and then drops between ovulation and fertilization. The more recent data on RNA synthesis during oogenesis and oocyte maturation in the mouse have been obtained using autoradiographic methods. It was shown that ^{3}H-uridine incorporation into the oocytes of small resting follicles in vivo is relatively low, but increases when the follicle and the oocyte begin to grow. RNA synthesis reaches a peak in oocytes of multilayered follicles (100-200 granulosa cells). At this developmental time the oocyte has attained its maximum size, while follicular growth continues, reaching its final size in preovulatory follicles (see review by *Pedersen* and *Peters*, 1968). With the appearance of follicular fluid and the formation of a definite antrum, RNA synthesis declines and ^{3}H-uridine uptake is very low or even absent in oocytes approaching maturity (*Oakberg*, 1967, 1968; *Moore* et al., 1974). Also, in full-grown oocytes of the rat and monkey, RNA synthesis is no longer detectable (*Baker* et al., 1969). Since the morphologic and qualitative pattern of development of the oocyte and follicle is very similar in all mammalian species, *Moore* and *Lintern-Moore* (1974) concluded that the pattern of RNA synthesis during oogenesis and maturation must also be very similar. RNA synthesis has been assayed by autoradiography also with oocytes in vitro, cultured in medium containing ^{3}H-uridine (*Bloom* and *Mukherjee*, 1972). From the results of these investigations it can be concluded that RNA is synthesized only before the breakdown of the germinal vesicle, but not in the mature oocyte. It is not possible, however, to conclude from these studies on the rate of RNA synthesis actually occurring during oogenesis, because the size and the state of the internal precursor pool is not known at different stages of oocyte development.

The mechanism by which RNA synthesis in mature oocytes is suppressed is unclear. A similar situation existing in mature amphibian oocytes is presumably due to a protein factor in the germinal vesicle which is able to inhibit the transcription of ribosomal genes (*Crippa*, 1970; *Crippa* et al., 1972). According to *Moore* et al. (1974), there is some indication that the antral fluid of the mammalian follicle contains a substance, which inhibits DNA dependent RNA polymerase, required for RNA synthesis. In fact, RNA polymerase activity is very low in the unfertilized mouse egg (*Siracusa*, 1973; *Siracusa* and *Vivarelli*, 1975).

The great bulk of RNA in a cell and also in the mammalian oocyte is ribosomal RNA. As a rough estimate one can say that the mammalian egg contains approximately 100 times the amount of rRNA compared to the adult cell. *Bachvarova* (1974) calculated the amounts of (28 S + 18 S) rRNA per mouse egg with 0.2 ng and tRNA plus 5 S-RNA with 0.16 ng per egg. *Young* et al. (1973) obtained a value for ribosomal RNA of 0.4 ng per egg, which is twice that found by *Bachvarova* (1974). The discrepancy between both determinations might be due to different experimental procedures. The synthesis of eukaryotic rRNA is manifest in the presence of a nucleolus, and nucleolar size and rRNA synthesis are correlated (see review by *Sirlin*, 1972). In accordance with this rule oocyte nucleolar size is positively and linearly correlated with RNA synthesis and the size increases during the oocyte growth period and stops when the oocyte has reached its maximal size and RNA synthesis declines (see review by *Chouinard*, 1971; *Moore* and *Lintern-Moore*, 1974; *Lintern-Moore* et al., 1974).

2. Does rDNA Amplification also Occur in Mammals?

The enormous amount of rRNA in the mammalian oocyte as compared to the adult cell gives rise to the question whether an amplification of ribosomal genes does occur during mammalian oogenesis as evidenced in other taxonomic orders (see above). Amplification should lead to an increase in the DNA content of the egg. *Alfert* (1950) and *Abramczuk* and *Sawicki* (1975), using the Feulgen method, found only the diploid quantity of DNA in the nuclei of fertilized mouse eggs. A corresponding result has been obtained in the rabbit oocyte (*van de Kerckhove*, 1959). However, *Olds* et al. (1973), extracting the nucleic acids by a modified Schmid-Tannhauser method, determined the DNA content of the unfertilized mouse egg with 23.8 pg. Mouse spermatozoa serving as controls yielded DNA amounts of 3.82 pg, which corresponds well with the DNA content of 8.4 pg in the diploid mouse hepatocyte evaluated by Feulgen cytophotometry (*Bachmann*, 1972). The unfertilized egg is diploid. Even if a "4N"-like figure must be taken into account, because the polar body may remain imprisoned in the confines of the zona pellucida, excess DNA should be present in the unfertilized egg. A much higher DNA content in the mouse unfertilized egg was found by *Reamer* (1963). According to this author the mouse oocyte contains 45.8 pg DNA at ovulation and 49.0 pg DNA after fertilization. In this connection the observations of *Stahl* et al. (1975) are of interest, demonstrating that in addition to three large nucleoli, micronucleoli, frequently numbering 15-20, are present in the diplotene stage of human oogenesis. The presence of micronucleoli might be interpreted that the human oocyte, and perhaps the mammalian oocyte in general, show an amplification of ribosomal genes. This amplification may likewise serve as templates in the synthesis of rRNA for use during early cleavages. Ribosomal RNA synthesis starts only at the four cell stage in the mouse and as late as at the morula to blastocyst transition in the rabbit.

3. Synthesis and Storage of Messenger RNA

It is remarkable that lampbrush chromosomes, which are assumed to represent the site of synthesis of mRNA for storage in invertebrates and lower vertebrates, are found in growing oocytes throughout the animal kingdom. Apparently, lampbrush chromosomes are almost as ancient in evolution as the process of oogenesis itself (see reviews by *Baker*, 1971; *Davidson* and *Hough*, 1972). Therefore, it seems justifiable to assume that in the growing mammalian oocyte, mRNA is synthesized for later use.

In the cytoplasm of oocytes of some mammalian species, highly ordered arrays of parallel chains or lattice-like structures have been described (*Weakley*, 1966, 1968; *Zamboni* and *Mastroianni*, 1966; *Schlafke* and *Enders*, 1967; *Burkholder* et al., 1971). These structures first appear during the follicular oocyte growth period and are fully expressed in the mature, preovulatory oocyte. However, they disappear completely during the early cleavage stages. Recently, *Burkholder* et al. (1971) reinvestigated the lattice-like structures of the mouse oocyte and postulated that they might represent the possible formation of maternally derived inactive ribosome-mRNA complexes. The activation of these complexes after fertilization might be caused by proteolytic enzymes that are known to be present in the mammalian egg (*Anderson*, 1972; *Stambaugh* and *Buckley*, 1972; *Gwatkin* et al., 1973; *Gwatkin* and *Williams*, 1974).

Bachvarova (1974) performed experiments recently in which ovulated mouse ova were labeled in vivo by exposure of growing oocytes to ^{3}H-adenosine and ^{3}H-uridine, respectively.

The precursors were injected into the ovarian bursa, ova were collected by superovulation 7-19 days after the injection. This period is known to be the major growth phase of the oocyte before ovulation. The labeled RNA from the ovulated eggs was analyzed by polyacrylamide – SDS electrophoresis. The ovaries from the same individuals were used as controls. The bulk of labeled RNA was found in the ribosomal and transfer RNA fractions. However, the electrophoretic profiles suggest a significant amount of labeled heterogeneous RNA (10-15% of the total incorporation) under and near the 18 S and 28 S peaks. Heterogeneous RNA is thought to be the precursor of polysomal mRNA (see reviews by *Darnell* et al., 1973; *Scherrer*, 1973).

Further evidence for the presence of mRNA in the mammalian oocyte comes from investigations dealing with the effect of the mushroom toxin α-amanitin on early cleavages of rabbit and mouse (*Manes*, 1973; *Golbus* et al., 1973). α-Amanitin is a specific inhibitor of RNA-polymerase II, an enzyme that is required for the synthesis of heterogeneous RNA (*Lindell* et al., 1970; *Jacob* et al., 1970; *Roeder* et al., 1975). Rabbit embryos initially exposed to the toxin while at the one cell stage can continue to cleave and develop for two to three cell generations, arresting at the eight cell stage. This observation implies that the first three cleavages in the rabbit are possible without the participation of embryonally coded mRNA. They must therefore be due to stored maternally derived mRNA in the oocyte. However, the arrest of the embryo at the eight cell stage indicates that gene activation in the rabbit embryo has to occur as early as at the one cell stage to support development after the eight cell stage. *Golbus* et al. (1973) obtained similar results with α-amanitin in the mouse. Therefore, it can be assumed that the egg cytoplasm, and the embryonic genome both contribute to allow for protein synthesis in early mammalian development.

Similar investigations with respect to the question as to what extent maternally transmitted mRNA governs development in mammals, have been performed in the mouse and rabbit with actinomycin D, a toxin that also inhibits transcription. However, the mammalian embryo has been found to be extraordinarily sensitive to this toxin, leading to an arrest of cleavage at low concentrations (*Mintz*, 1964; *Thomson* and *Biggers*, 1966; *Skalko* and *Morse*, 1969; *Tasca* and *Hillman*, 1970; *Monesi* et al., 1970; *Golbus* et al., 1973; *Manes*, 1973). On the other hand, protein synthesis in the early embryo is largely resistant to prolonged treatment with actinomycin D (*Manes* amd *Daniel*, 1969; *Monesi* et al., 1970; *Tasca* and *Hillman*, 1970), also indicating that maternally derived stable RNA molecules are present in the zygote which are translated during early cleavages. The inhibitory effect of actinomycin D on early mammalian development is assumed to be due to "side effects" of the drug (*Manes* and *Daniel*, 1969; *Wolf* and *Engel*, 1972; *Manes*, 1973), as described by *Revel* et al. (1964), *Honig* and *Rabinovitz* (1965), *Lazslo* et al. (1966), *Soeiro* and *Amos* (1966), *Pastan* and *Friedman* (1968), and *Steinberg* et al. (1975).

It is known that most mRNA molecules contain polynucleotide sequences rich in adenylic acid [poly (A)] covalently linked to the 3' end of the molecule (*Lim* and *Canellakis*, 1970; *Burr* and *Lingrel*, 1971; *Darnell* et al., 1971; *Edmonds* et al., 1971; *Lee* et al., 1971). Poly (A) sequences are not present in histone mRNA (*Adesnik* and *Darnell*, 1972; *Greenberg* and *Perry*, 1972), and in ribosomal or transfer RNA (*Edmonds* et al., 1971; *Lee* et al., 1971; *Greenberg* and *Perry*, 1972; *Sheldon* et al., 1972). A recent report by*Milcarek* et al. (1974) indicates that as much as 30% of the HeLa cell mRNA lacks poly (A) sequences.

Nevertheless, the poly (A) tract may be considered a useful marker for detecting putative mRNA in the mammalian oocyte, as exemplified with oocytes and embryos of the sea urchin and *X. laevis* (*Slater* et al., 1973; *Wilt*, 1973; *Mescher* and *Humphreys*, 1974; *Rosbash* and *Ford*, 1974) and in rabbit preimplantation development (*Schultz* et al., 1973; *Schultz*, 1973, 1974). Performing hybridization experiments with ^{3}H-polyuridylic acid and unlabeled RNA from unfertilized and fertilized rabbit eggs, *Schultz* (1975) recently presented evidence for the presence of poly (A)-containing RNA species in the rabbit egg, which should have been synthesized during oogenesis. The proportion of poly (A)-containing RNA in the rabbit egg was determined with 0.25% of the total RNA amount, which is somewhat higher than that reported for the sea urchin egg: 0.033% (*Mescher* and *Humphreys*, 1974), 0.045% (*Wilt*, 1973) and 0.186% (*Slater* et al., 1973), but similar to values reported for mammalian somatic cells. In the rabbit egg, poly (A)-containing RNA is detectable in the nuclear, ribosomal and subribosomal fractions. However, the majority of the adenylated RNA is localized in the ribosomal fraction, indicating that these are maternal mRNAs associated with polysomes (*Schultz*, 1975). According to the investigations of *Manes* (1973), who exposed fertilized rabbit eggs to α-amanitin, these maternally derived ribonucleoprotein complexes account for 50% of the protein synthesis during early cleavages. Newly synthesized mRNA molecules in the rabbit embryo have been found to occur as early as the 16 cell stage (*Schultz* et al., 1973). While the existence of mRNA in the mammalian oocyte seems to be established, the more important questions, perhaps, what types of mRNA present in the newly ovulated egg, and when are they used during early development, still remain open.

4. The Proteins in the Mammalian Egg

The other important type of reserve substance in the mammalian egg are the proteins. A number of studies were published on protein synthesis during oogenesis and oocyte maturation and on the total protein amount in the unfertilized egg of mammals (*Alfert*, 1950; *Hedberg*, 1953; *Flax*, 1953; *Lin*, 1956; *Greenwald* and *Everett*, 1959; *Roversi* and *Silvestrini*, 1963; *Mintz*, 1964; *Loewenstein* and *Cohen*, 1964; *Brinster*, 1967d, 1971c; *Manes* and *Daniel*, 1969; *Baker* et al., 1969; *Stern* et al., 1972; *Petzoldt* et al., 1972; *Cross* and *Brinster*, 1974; *Epstein* and *Smith*, 1974; *van Blerkom* and *Manes*, 1974; *van Blerkom* and *Brockway*, 1975). The protein content of the unfertilized mouse egg has been determined with 20 ng (*Loewenstein* and *Cohen*, 1964) or 27.8 ng (*Brinster*, 1967a) and with 100 ng in the unfertilized rabbit egg (*Brinster*, 1971a). No reliable data exist till now on the amount of protein nor on the synthesis of protein in the mammalian oocyte. *Flax* (1953) found that Fast-Green stainable basic protein in the cytoplasm of the mouse oocyte shows an increase during oogenesis which is almost linear with increasing cytoplasmic volume. A peak is reached shortly before maximum oocyte size is attained, followed by a slow decline. Most information on protein synthesis in oocytes comes from radioautographs of sectioned ovaries after injections of different labeled amino acids. The results obtained indicate that protein synthesis is performed in oocytes at all stages of follicular development, but not in the unfertilized egg. However, the value of autoradiographic investigations of sectioned ovaries is limited, because it is impossible to distinguish between protein synthesized by the oocyte itself, and that synthesized by other cells and transferred into the oocyte. It has been shown that maternal plasma molecules and administered foreign proteins can be incorporated nearly intact from the blood into the oocyte (see review by *Glass*, 1971). Recently, *Stern* et al. (1972) and *Cross* and *Brinster* (1974) have overcome these difficulties by incubating mouse oocytes free from all cellular investments in an in vitro system containing both labeled valine and leucine. Comparing total leucine uptake of oocytes at the dictyate, metaphase I, and metaphase II stages of maturation, the highest leucine uptake was found in oocytes at metaphase I. Since the internal amino acid pool during oocyte development is not known, the only conclusion from these results is that mouse oocytes are capable of protein synthesis.

In the unfertilized mammalian egg, different enzymes have been detected (see reviews by *Biggers*, 1971; *Wolf* and *Engel*, 1972; *Church* and *Schultz*, 1974; *Epstein*, this volume, and Table 1). Enzyme activities were determined mainly in the mouse, rat, and rabbit. It has been observed that the activity for a definite enzyme in the unfertilized egg differs from species to species. However, up to now, in growing oocytes only the activities of the enzymes lactate dehydrogenase (LDH; E.C. 1.1.1.27) and glucose-6-phosphate dehydro-

Table 1. Enzymes demonstrated to be present in the mammalian egg

Enzymes	References
Pyruvate carboxylase	*Quinn* and *Wales* (1971)
NADP-malate dehydrogenase	*Quinn* and *Wales* (1971)
NAD-glycohydrolase	*Streffer* and *van Beuningen* (1974)
6-Phosphogluconate dehydrogenase	*Brinster* (1971a)
Glucose-6-phosphate dehydrogenase	*Brinster* (1966a, 1970a)
RNA-polymerase	*Siracusa* (1973): *Siracusa* and *Vivarelli* (1975)
Phosphoglucose isomerase	*Chapman* et al. (1971)
Lactate dehydrogenase	*Brinster* (1965a, 1967a, b, c, 1968b)
Fructose-1,6-diphosphate aldolase	*Epstein* et al. (1969)
NAD-malate dehydrogenase	*Brinster* (1966b); *Epstein* et al. (1969)
Guanine deaminase	*Epstein* et al. (1971)
Hypoxanthine-guanine-phosphorybosyl-transferase	*Epstein* (1972)
Adenine-phosphoribosyl-transferase	*Epstein* (1970)
Hexokinase	*Brinster* (1968a)
Glutamic dehydrogenase	*Moore* and *Brinster* (1970)
Phosphofructokinase	*Brinster* (1971b)
Glycogen synthetase	*Stern* (1970)
Aspartate aminotransferase	*Moore* and *Brinster* (1970)
NADP-isocitrate dehydrogenase	*Donahue* and *Stern* (1970)
Uridine kinase	*Daentl* and *Epstein* (1971)
Triose phosphate isomerase	*Quinn* and *Kozak* (1975)
Glyceraldehyd-3-phosphate dehydrogenase	*Quinn* and *Kozak* (1975)
Phosphoglyceromutase	*Quinn* and *Kozak* (1975)
Phosphoglycerokinase	*Quinn* and *Kozak* (1975)

genase (G-6-PD; E.C. 1.1.1.49) have been studied (*Mangia* and *Epstein*, 1975). Total activity of both enzymes was found to increase continuously as the mouse oocyte grows. However, in oocytes which have reached more than 80-85 μ in diameter, the specific activity of both enzymes declines. This result suggested to the authors that enzyme synthesis is completed at this growth stage. The LDH activity course in growing mouse oocytes as determined by *Mangia* and *Epstein* (1975) is confirmed by our own measurements, using oocytes of different diameters from mice between 7 and 21 days of age. However, according to our results, LDH activity in the ovarian oocyte is not only dependent on the size of the oocyte but also on the age of the animal (unpublished results).

As shown in Table 1, the activities of some enzymes, known to be X-linked (*Ohno*, 1973) have been determined in unfertilized eggs. *Epstein* (1969, 1972) found that in eggs of XX-mice the activities of G-6-PD and hypoxanthine-guanine phosphoribosyltransferase (HGPRT; E.C. 2.4.2.8) is twice as high as in eggs of XO-mice. Corresponding results have been obtained with the enzyme phosphoglycerokinase (PGK; E.C. 2.7.2.3) (*Quinn* and *Kozak*, 1975). These results support *Ohno*'s view (*Ohno*, 1967) that in the mammalian oocyte prior to the reductional division, both X-chromosomes are genetically active. Recently, *Gartler* et al. (1972) demonstrated in a G-6-PD heterozygote woman a three-banded electrophoretic pattern for this enzyme in mature oocytes, while in other cell types the heteromeric G-6-PD band was lacking. The presence of the heteromeric G-6-PD band in fetal oocytes of a G-6-PD heterozygote embryo is to be interpreted that the female germ cells are never subject to X-inactivation (*Gartler* et al., 1973).

The enzyme which has been investigated most extensively in the mammalian oocytes is lactate dehydrogenase (LDH). This enzyme catalyzes the conversion of lactate to pyruvate.

Table 2. Lactate dehydrogenase activity in oocytes of mammals.
Values are means ± S.E.M.

Species	Enzyme activity [NADH oxidized/oocyte/h (moles x 10^9)]	
Human	1.30 ± 0.11	*Brinster* (1967a)
Squirrel monkey	1.50 ± 0.16	
Rhesus monkey	6.32 ± 0.34	
Mouse	47.69 ± 1.78	
Rabbit	1.99 ± 0.13	
Rat	17.66 ± 0.31	*Brinster* (1968)
Golden hamster	5.80 ± 0.43	
Guinea pig	3.97 ± 0.16	
Cow	12.35 ± 0.48	
Ferret	6.65 ± 0.81	
Chinese hamster	18.24 ± 1.08	Own published data
Mus musculus (wild)	35.88 ± 2.64	
Microtus arvalis	4.08 ± 0.18	
Clethrionomys glareolus	10.32 ± 1.14	
Apodemus sylvaticus	13.74 ± 1.13	
Meriones unguiculatus	3.12 ± 0.42	

It has been shown that LDH comprises about 5% of the total protein in the mouse ovum and that its activity in the oocyte is 10 times higher than in any other tissue in the mouse (*Brinster*, 1965a). However, a comparison between the LDH activity in mature oocytes of different species reveals great variations. Whereas the laboratory mouse shows the highest activity, the human and rabbit display the lowest found so far (*Brinster*, 1967a, 1968b). However, even different mouse species differ in their LDH activity in oocytes (Table 2).

The LDH activity in the mouse ovum consists electrophoretically of only a single band, identified as LDH1 (*Rapola* and *Koskimies*, 1967; *Auerbach* and *Brinster*, 1967). This result has been corroborated by immunologic studies (*Spielmann* et al., 1974). We investigated the LDH isozyme pattern by microdisc electrophoresis in full-grown oocytes of 16 mammalian species belonging to five different taxonomic orders. The results obtained point to the existence of two types of oocytes in mammals: while in the order *Rodentia* only LDH1 is demonstrable, LDH isozymes formed of α- and β-subunits are present in the oocytes of species of the orders *Lagomorpha, Carnivora, Artiodactyla*, and in man. However, LDH1 is the most intense isozyme also in these species (*Engel* and *Kreutz*, 1973; *Engel* et al., 1975).

The significance of the high LDH activity in the mouse eggs is unclear, particularly since mouse zygotes are known to be unable to convert lactate to pyruvate. Pyruvate has been found to be the central energy source for the early developmental stages of mammalian embryos. Under in vitro conditions, in the oocyte and fertilized mouse egg only pyruvate and oxalacetate, but not lactate would allow development, while in the two cell embryo, pyruvate, oxalacetate, phosphoenolpyruvate, and lactate support development to the blastocyst (*Whitten*, 1957; *Brinster*, 1965b, 1970b; *Biggers* et al., 1967; *Whittingham*, 1969). Therefore, it has been suggested that LDH activity in the ovum is a carry-over of high activity necessary during oogenesis. *Zeilmaker* et al. (1972), *Sorensen* (1972), and *Cross* and *Brinster* (1973) assumed that LDH cannot function due to a very low level of NAD in the mouse zygote. These authors have found that mouse zygotes cleave successfully in vitro in the presence of lactate *and* NAD. It was observed, however, that the NAD content in mouse ova or fertilized eggs is even higher than in the following preimplantation stages (*Kuwahara* and *Chaykin*, 1973; *Streffer* and *van Beuningen*, 1974).

In contrast to the mouse, development of the rat zygote to the two cell stage can occur if lactate is present in the media (*Spielmann*, 1975). Therefore, despite the uncertainty in the mouse, LDH can be regarded as an important enzyme for early mammalian embryos and oocytes.

5. Inactive Enzyme Molecules and their Activation

There is some evidence that in the mammalian oocyte and egg, besides active enzyme molecules, inactive enzyme molecules are present. While in rat ova, only LDH1 is normally demonstrable (*Cornette* et al., 1967; *Engel* and *Kreutz*, 1973; *Engel* and *Petzoldt*, 1973), treatment of the eggs with Triton X-100 (0.5%) results in the appearance of additional isozymes, namely LDH2 and LDH5 (*Poznakhirkina* et al., 1975). It has been reported that in the rabbit, contrary to the situation in mouse and rat, LDH isozymes formed of α- and β-subunits are present in ova as well as in zygotes (*Brinster*, 1973a; *Engel* et al., 1975). Oocytes in the germinal vesicle stage exhibit only LDH1, but they change their LDH-isozyme pattern during the gonadotropin-induced progression from late prophase of the I. meiotic division to metaphase of the II. meiotic division, when ovulation occurs. Eight h after the hCG injection, the LDH isozyme pattern in hormonally primed oocytes corresponds to that of ovulated eggs. Because gene activity during oogenesis is restricted to the lampbrush stage of oocyte growth, the LDH isozymes newly appearing during oocyte maturation in the rabbit should not be due to transcriptional activity. Evidence has been presented that the new isozymes derive from inactive precursors (*Engel* et al., 1975).

The results of recent experiments in our laboratory indicate that inactive LDH enzyme molecules are also stored in the mouse. NMRI mice, 21 days old, were primed with PMS (Anteron, SCHERING AG, Berlin) and injected with hCG (Primogonyl, SCHERING AG, Berlin) 48 h later. Mice, 21 days old, were used for the experiments because of the large numbers of follicles capable of stimulation at this developmental stage (*Gates*, 1971). From the ovaries, only oocytes with a diameter of 75-100 μ were collected. The preparation and the handling of the oocytes were performed according to *Brinster* (1968b). The cumulus cells were removed by careful mechanical treatment. LDH activity measurements from single oocytes were performed as described by *Brinster* (1968b). Oocytes from hormonally untreated 21-day-old mice exhibited an LDH activity of about 4.95 x 10^{-8} moles of substrate converted per oocyte per hour, remaining constant in oocytes during the PMS treatment period and during the first 3 h after the hCG injection (Table 3). It is known that oocytes during this developmental stage are in the germinal vesicle stage being morphologically indistinguishable from those obtained from nonhormonally stimulated animals (*Calarco* et al., 1972). After the hCG-injection, the oocytes rapidly pass the remaining stages of meiosis, reaching metaphase I 4-5 h and metaphase II 12-14 h later (*Edwards* and *Gates*, 1959; *Donahue*, 1972). Therefore, in animals primed with hCG longer than 3 h, only oocytes without a germinal vesicle were used for LDH measurements. As shown in Table 3, an increase in LDH activity is observed in oocytes 5 h after the hCG injection, remaining nearly constant thereafter. Enzyme activity does not differ significantly between oocytes 5 h after the hCG treatment and ovulated eggs. Because of the high LDH activity in metaphase I oocytes compared to the germinal vesicle stage oocytes, we assume that enzyme activity increase precedes or even coincides with the building up of the meiotic metaphase I.

Table 3. Lactate dehydrogenase activity [(NADH oxidized/oocyte/h (moles x 10^9)] during oocyte maturation in 21-day-old mice. Values are means ± S.E.M.; values in parentheses are number of determinations

Hormonally untreated mice	49.50 ± 2.28 (40)
24 h following PMS	48.36 ± 4.26 (17)
48 h following PMS	51.42 ± 3.84 (18)
2 h following hCG	51.20 ± 3.20 (13)
3 h following hCG	51.00 ± 3.60 (16)
5 h following hCG	59.58 ± 2.70 (11)
10 h following hCG	60.26 ± 2.92 (10)
15 h following hCG	
ovulated eggs	61.50 ± 3.6 (29)
ovarian oocytes	49.56 ± 3.84 (14)

This LDH activity increase during oocyte maturation in the mouse cannot be due to transcriptional activity, because transcription stops in the antrum stage of oogenesis. We incubated germinal vesicle stage oocytes with trypsin (0.025%) and pronase (0.25%) for 3 and 1 minutes respectively, and found that LDH activity increases similar to that in metaphase I oocytes; hyaluronidase treatment (1000 IU/ml) had no effect on LDH activity. From these results it is concluded that the LDH activity increase during oocyte maturation might be due to the activation of inactive LDH enzyme molecules or stored inactive mRNA molecules for LDH. However, because of the very short incubation time involved before getting maximal response, we consider the mRNA hypothesis to be highly unlikely. This view was supported by LDH activity measurements in germinal vesicle stage oocytes after simultaneous incubation with trypsin and cycloheximide. Cycloheximide, a drug which inhibits translation of mRNA, did not antagonize the trypsin induced effect.

We, therefore, assume that the LDH activity increase in germinal vesicle stage oocytes after trpysin or pronase treatment is due to activation of inactive protein-coded enzyme molecules. In vivo, the activation might be effected by proteases that become liberated in the oocyte from lysosomes or other vesicles around the time of metaphase I. A more detailed report of these results, as well as on similar results with the enzyme phosphoglucose isomerase will be published elsewhere.

III. Final Remarks

The oocytes of invertebrates, lower vertebrates, and mammals are endowed with maternally derived cytoplasmic storage substances, including mRNA. These reserves are built up during oogenesis and used during early development. In our opinion, the eggs of the sea urchin, of *X. laevis*, and the mammal egg differ from each other with respect to storage only quantitatively. This might be the cause for the observation that the mammalian genome has to be activated as early as in the one cell (= zygote) stage. However, in mammals as in invertebrates or lower vertebrates, development prior to gastrulation should depend on both, stored and new gene products. A very interesting question in this connection is whether templates might exist in the egg cytoplasm, having a role in recognizing individual gene loci within the zygotic nuclei and activating them at appropriate stages of early development. Evidence for the existence of such regulatory substances in the egg comes from studies on gene expression in organismal hybrids. In interspecific hybrids between rather remotely related species of birds and fish, in most cases only the maternally derived allele of the structural gene locus was found to become activated according to the time shedule of the maternal species, while the activation of the paternally derived allele is delayed or even absent (*Hitzeroth* et al., 1968; *Ohno,* 1969; *Whitt* et al., 1973). Similar investigations in mammals using mice strains that differ from each other in the electrophoretic position for a definite enzyme, have scarcely been performed till now (*Donahue* and *Stern*, 1970; *Chapman* et al., 1971; *Engel* and *Wolf*, 1971; *Wolf* and *Engel*, 1972; *Brinster*, 1973b; *Engel*, 1973). They indicate rather synchronous activation of the parental alleles. However, the possibility that regulatory substances are present in the mammalian egg is supported by the observation that the triggering and control of DNA- and RNA-synthesis after fertilization might be under the control of the egg cytoplasm (*Bernstein* and *Mukherjee*, 1973; *Abramczuk* and *Sawicki*, 1975).

Acknowledgment: We are greatly indebted to Professor *U. Wolf* and Dr. *J. Schmidtke* for critical discussions and for reading the manuscript. This work was supported by the Deutsche Forschungsgemeinschaft (SFB 46).

References

Abramczuk, J., Sawicki, W.: Pronuclear synthesis of DNA in fertilized and parthenogenetically activated mouse eggs. Exp. Cell Res. 92, 361-372 (1975).

Adesnik, M., Darnell, J.E.: Biogenesis and characterization of histone messenger RNA in HeLa cells. J. molec. Biol. 67, 397-406 (1972).

Alfert, M.: A cytochemical study of oogenesis and cleavage in the mouse. J. cell. comp. Physiol. 36, 381-410 (1950).

Anderson, E.: The localization of acid phosphatase and the uptake of horseradish peroxidase in the oocyte and follicle cells of mammals. In: Oogenesis. Baltimore: University Park Press 1972.

Auerbach, S., Brinster, R.L.: Lactate dehydrogenase isozymes in early mouse embryo. Exp. Cell Res. 46, 89-92 (1967).

Austin, C.R.: The mammalian egg. Illinois: Thomas Springfield 1961.

Austin, C.R., Braden, W.H.: The distribution of nucleic acids in rat eggs in fertilization and early segmentation. I. Studies on living eggs by ultraviolet microscopy. Aust. J. biol. Sci. 6, 324-333 (1953).

Bachmann, K.: Genome size in mammals. Chromosoma 37, 85-93 (1972).

Bachvarova, R.: Incorporation of tritiated adenosine into mouse ovum RNA. Develop. Biol. 40, 52-58 (1974).

Baker, T.G.: Electron microscopy of the primary and secondary oocyte. Advanc. Biosci. 6, 8-27 (1971).

Baker, T.G., Beaumont, H.M., Franchi, L.L.: The uptake of tritiated uridine and phenylalanine by the ovaries of rats and monkeys. J. Cell Sci. 4, 655-675 (1969).

Balhorn, R., Tanphaichitr, N., Chalkley, R., Granner, D.K.: The effect of inhibition of desoxyribonucleic acid synthesis on histone phosphorylation. Biochemistry 12, 5146-5150 (1973).

Beatty, R.A.: Parthenogenesis and heteroploidy in the mammalian egg. In: Oogenesis. Baltimore: University Park Press 1972.

Beauregard, L.J., Ghosal, S.K., Lamarche, P.H.: R.N.A. synthetic activity of human oocyte. Lancet 1968 II, 1039.

Benson, P.F.: Regulation of genetic expression. In: The Biochemistry of Development. Philadelphia: J.B. Lippincott Co. 1971.

Bernstein, R.M., Mukherjee, B.B.: Cytoplasmic control of nuclear activity in preimplantation mouse embryos. Develop. Biol. 34, 47-65 (1973).

Biggers, J.D.: Metabolism of the oocyte. In: Oogenesis. Baltimore: University Park Press 1972.

Biggers, J.D., Whittingham, D.G., Donahue, R.P.: The pattern of energy metabolism in the mouse oocyte and zygote. Proc. nat. Acad. Sci. (Wash.) 58, 560-567 (1967).

Blerkom, J. van, Brockway, G.A.: Qualitative patterns of protein synthesis in the preimplantation mouse embryo. I. Normal pregnancy. Develop. Biol. 44, 148-157 (1975).

Blerkom, J. van, Manes, C.: Development of preimplantation rabbit embryos in vivo and in vitro. II. A comparison of qualitative aspects of protein synthesis. Develop. Biol. 40, 40-51 (1974).

Bloom, A.M., Mukherjee, B.B.: RNA synthesis in maturing mouse oocytes. Exp. Cell Res. 74, 577-582 (1972).

Bomsel-Helmreich, O.: Fate of heteroploid embryos. Advanc. Biosci. 6, 381-403 (1971).

Brachet, J., Denis, H., de Vitry, F.: The effects of actinomycin D and puromycin on morphogenesis in amphibian eggs and Acetabularia mediterranea. Develop. Biol. 9, 398-434 (1964).

Briggs, R.W., Green, E., King, T.J.: An investigation of the capacity for cleavage and differentiation in Rana pipiens eggs lacking "functional" chromosomes. J. exp. Zool. 116, 455-468 (1951).

Briggs, R.W., Signoret, J., Humphrey, R.R.: Transplantation of nuclei of various cell types from neurulae of the mexican axolotl (Ambystoma mexicanum). Develop. Biol. 10, 233-246 (1964).

Brinster, R.L.: Lactic dehydrogenase activity in the preimplanted mouse embryo. Biochim. biophys. Acta (Amst.) 110, 439-441 (1965a).

Brinster, R.L.: Studies on the development of mouse embryos in vitro. II. The effect of energy source. J. exp. Zool. 158, 59-68 (1965b).

Brinster, R.L.: Glucose-6-phosphate dehydrogenase activity in the preimplantation mouse embryo. Biochem. J. 101, 161-163 (1966a).

Brinster, R.L.: Malic dehydrogenase activity in the preimplantation mouse embryo. Exp. Cell Res. 43, 131-135 (1966b).

Brinster, R.L.: Lactate dehydrogenase activity in human, squirrel monkeys and Rhesus monkey oocytes. Exp. Cell Res. 48, 643-646 (1967a).

Brinster, R.L.: Lactate dehydrogenase activity in the preimplantation rabbit embryo. Biochim. biophys. Acta (Amst.) 148, 298-300 (1967b).

Brinster, R.L.: Lactic dehydrogenase activity in preimplantation rat embryo. Nature (Lond.) 214, 1246-1247 (1967c).

Brinster, R.L.: Protein content of the mouse embryo during the first five days of development. J. Reprod. Fertil. 13, 413-420 (1967d).

Brinster, R.L.: Hexokinase activity in the preimplantation mouse embryo. Enzymologia 34, 304-308 (1968a).

Brinster, R.L.: Lactate dehydrogenase activity in the oocyte of mammals. J. Reprod. Fertil. 17, 139-146 (1968b).

Brinster, R.L.: Glucose-6-phosphate dehydrogenase activity in the early rabbit and mouse embryo. Biochem. Genet. 4, 669-676 (1970a).

Brinster, R.L.: Metabolism of the ovum between conception and nidation. In: Mammalian Reproduction. Berlin-Heidelberg-New York: Springer 1970b.

Brinster, R.L.: Activity of 6-phosphogluconate dehydrogenase in the preimplantation mouse and rabbit embryo. Experientia (Basel) 27, 371-372 (1971a).

Brinster, R.L.: Phosphofructokinase activity in the preimplantation mouse embryo. Wilhelm Roux' Arch. Entwickl.-Mech. Org. 166, 300-302 (1971b).

Brinster, R.L.: Uptake and incorporation of amino acids by the preimplantation mouse embryo. J. Reprod. Fertil. 27, 329-338 (1971c).

Brinster, R.L.: Lactate dehydrogenase isozymes in the preimplantation rabbit embryo. Biochem. Genet. 9, 229-234 (1973a).

Brinster, R.L.: Parental glucose phosphate isomerase activity in three-day mouse embryos. Biochem. Genet. 9, 187-191 (1973b).

Brown, D.D.: The genes for ribosomal RNA and their transcription during amphibian development. Curr. Top. develop. Biol. 2, 47-64 (1967).

Brown, D.D., Dawid, I.B.: Specific gene amplification in oocytes. Science 160, 272-280 (1968).

Brown, D.D., Dawid, I.B.: Developmental genetics. Ann. Rev. Genet. 3, 127-154 (1969).

Brown, D.D., Gurdon, J.B.: Absence of ribosomal RNA synthesis in annucleolate mutant of Xenopus laevis. Proc. nat. Acad. Sci. (Wash.) 51, 139-146 (1964).

Brown, D., Littna, E.: RNA synthesis during the development of Xenopus laevis, the South African clawed toad. J. molec. Biol. 8, 669-687 (1964).

Burkholder, G.D., Comings, D.E., Okada, T.A.: A storage form of ribosomes in mouse oocytes. Exp. Cell Res. 69, 361-371 (1971).

Burr, H., Lingrel, J.B.: Poly A sequences at the 3'-Termini of rabbit globin mRNAs. Nature New Biol. (Lond.) 233, 41-43 (1971).

Calarco, P., Donahue, R.P., Szöllosi, D.: Germinal vesicle breakdown in the mouse oocyte. J. Cell Sci. 10, 369-385 (1972).

Callan, H.G.: Chromosomes and nucleoli of the axolotl, Ambystoma mexicanum. J. Cell Sci. 1, 85-108 (1966).

Chapman, V.M., Whitten, W.K., Ruddle, F.H.: Expression of paternal glucose phosphate isomerase-1 (Gpi-1) in preimplantation stages of mouse embryos. Develop. Biol. 26, 153-158 (1971).

Chouinard, L.A.: A light- and electron-microscope study of the nucleolus during growth of the oocyte in the prepubertal mouse. J. Cell Sci. 9, 637-663 (1971).

Church, R.B., Schultz, G.A.: Differential gene activity in the pre- and postimplantation mammalian embryo. Curr. Top. develop. Biol. 8, 179-202 (1974).

Clegg, K.B., Denny, P.C.: Synthesis of rabbit globin in a cell-free protein synthesis system utilizing sea urchin egg and zygote ribosomes. Develop. Biol. 37, 263-272 (1974).

Cornette, J.C., Pharris, B.B., Duncan, G.W.: Lactic dehydrogenase isozymes in the ovum and embryo of the rat. Physiologist 10, 146 (1967).

Crane, C.M., Villee, C.: The synthesis of nuclear histones in early embryogenesis. J. biol. Chem. 246, 719-723 (1971).

Crippa, M.: Regulatory factor for the transcription of the ribosomal genes in amphibian oocytes. Nature (Lond.) 227, 1138-1140 (1970).

Crippa, M., Davidson, E.H., Mirsky, A.E.: Persistence in early amphibian embryos of informational RNA's from the lampbrush chromosome stage of oögenesis. Proc. nat. Acad. Sci. (Wash.) 57, 885-892 (1967).

Crippa, M., Gross, P.R.: Maternal and embryonic contributions to the functional messenger RNA of early development. Proc. nat. Acad. Sci. (Wash.) 62, 120-127 (1969).

Crippa, M., Tocchini-Valentini, G.P., Andronico, F.: Regulation of ribosomal RNA synthesis during oogenesis of Xenopus laevis. In: Oogenesis. Baltimore: University Park Press 1972.

Cross, P.C., Brinster, R.L.: The sensitivity of one cell mouse embryo to pyruvate and lactate. Exp. Cell Res. 77, 57-62 (1973).

Cross, P.C., Brinster, R.L.: Leucine uptake and incorporation at three stages of mouse oocyte maturation. Exp. Cell Res. 86, 43-46 (1974).

Daentl., D.L., Epstein, C.J.: Developmental interrelationships of uridine uptake, nucleotide formation and incorporation into RNA by early mammalian embryos. Develop. Biol. 24, 428-442 (1971).

Darnell, J.E., Jelinek, W.R., Molloy, G.R.: Biogenesis of mRNA: Genetic regulation in mammalian cells. Science 181, 1215-1221 (1973).

Darnell, J.E., Philipson, L., Wall, R., Adesnik, M.: Polyadenylic acid sequences: Role in conversion of nuclear RNA into messenger RNA. Science 174, 507-510 (1971).

Davidson, E.H.: Gene Activity in Early Development. New York-London: Academic Press 1968.

Davidson, E.H., Hough, B.R.: Genetic information in oocyte RNA. J. molec. Biol. 56, 491-506 (1971).

Davidson, E.H., Hough, B.R.: Utilization of genetic information during oogenesis. In: Oogenesis. Baltimore: University Park Press 1972.

Davidson, E.H., Mirsky, A.E.: Gene activity in oogenesis. Brookhaven Symp. Biol. 18, 77-98 (1965).

Deane, H.W.: Histochemical observations on the ovary and oviduct of the albino rat during the oestrus cycle. Amer. J. Anat. 91, 363-413 (1952).

Denny, P.C., Tyler, A.: Activation of protein biosynthesis in non-nucleate fragments of sea urchin eggs. Biochem. biophys. Res. Commun. 14, 245-249 (1964).

Donahue, R.P.: The relation of oocyte maturation to ovulation in mammals. In: Oogenesis. Baltimore: University Park Press 1972.

Donahue, R.P., Stern, S.: Isocitrate dehydrogenase in mouse embryos: activity and electrophoretic variation. J. Reprod. Fertil. 22, 575-577 (1970).

Ecker, R.E., Smith, L.D.: Protein synthesis in amphibian oocytes and early embryos. Develop. Biol. 18, 232-249 (1968).

Edmonds, M., Vaughan, M.H., Nakazoto, H.: Polyadenylic sequences in the heterogenous nuclear RNA and rapidly labeled polyribosomal RNA of HeLa cells. Possible evidence for a precursor. Proc. nat. Acad. Sci. (Wash.) 68, 1336-1340 (1971).

Edwards, R.G., Gates, A.H.: Timing of the stages of the maturation divisions, ovulation, fertilization and the first cleavage of eggs of adult mice. J. Endocr. 18, 292-304 (1959).

Engel, W.: Onset of synthesis of mitochondrial enzymes during mouse development. Humangenetik 20, 133-140 (1973).

Engel, W., Franke, W., Petzoldt, U.: Isozymes as genetic markers in early mammalian development. In: Isozymes III. Developmental Biology. New York: Academic Press 1975.

Engel, W., Kreutz, R.: Lactate dehydrogenase isoenzymes in the mammalian egg: investigations by micro-disc-electrophoresis in 15 species of the orders Rodentia, Lagomorpha, Artiodactyla and in man. Humangenetik 19, 253-260 (1973).

Engel, W., Petzoldt, U.: Early developmental changes of the lactate dehydrogenase isoenzyme pattern in mouse, rat, Guinea-pig, Syrian hamster and rabbit. Humangenetik 20, 125-131 (1973).

Engel, W., Wolf, U.: Synchronous activation of the alleles coding for the S-form of the NADP-dependent malate dehydrogenase during mouse embryogenesis. Humangenetik 12, 162-166 (1971).

Epel, D.: Protein synthesis in sea urchin eggs: A "late" response to fertilization. Proc. nat. Acad. Sci. (Wash.) 57, 899-906 (1967).

Epstein, C.J.: Mammalian oocytes: X-chromosome activity. Science 163, 1078–1079 (1969).

Epstein, C.J.: Phosphoribosyltransferase activity during early mammalian development. J. biol. Chem. 245, 3289-3294 (1970).

Epstein, C.J.: Expression of the mammalian X-chromosome before and after fertilization. Science 175, 1467-1468 (1972).

Epstein, C.J., Daentl., D.L., Smith, S.A., Kwok, L.W.: Guanine metabolism in preimplantation mouse embryos. Biol. Reprod. 5, 308-313 (1971).

Epstein, C.J., Smith, S.A.: Electrophoretic analysis of proteins synthesized by preimplantation mouse embryos. Develop. Biol. 40, 233-244 (1974).

Epstein, C.J., Wagienka, E.A., Smith, C.W.: Biochemical development of preimplantation mouse embryos: in vivo activities of fructose 1,6-diphosphate aldolase, glucose-6-phosphate dehydrogenase, malate dehydrogenase, and lactate dehydrogenase. Biochem. Genet. 3, 271-281 (1969).

Evans, D., Birnstiel, M.L.: Localization of amplified ribosomal DNA in the oocyte of Xenopus laevis. Biochim. biophys. Acta (Amst.) 166, 274-276 (1968).

Farquhar, M.N., McCarthy, B.J.: Histone m-RNA in eggs and embryo of Strongylocentrotus purpuratus. Biochem. biophys. Res. Commun. 53, 515-522 (1973).

Flax, M.H.: Ribose nucleic acid and protein during oogenesis and early embryonic development in the mouse. Ph.D. Thesis, Columbia University (1953).

Ford, C.E.: Gross Genome unbalance in mouse spermatozoa: does it influence the capacity to fertilize? Proc. Int. Symp. Genet. Spermatozoon, Edinburgh, 16-20 Aug. 1971.

Ford, P.J.: Ribonucleic acid synthesis during oogenesis in Xenopus laevis. In: Oogenesis. Baltimore: University Park Press 1972.

Fry, B.J., Gross, P.R.: Patterns and rates of protein synthesis in sea urchin embryos. Develop. Biol. 21, 105-124 (1970).

Gall, J.G.: Differential synthesis of the genes for ribosomal RNA during amphibian oogenesis. Proc. nat. Acad. Sci. (Wash.) 60, 553-560 (1968).

Gall, J.G.: The genes for ribosomal RNA during oogenesis. Genetics 61, 121-132 (1969).

Gambino, R., Metafora, S., Felicetti, L., Raisman, J.: Properties of ribosomal salt wash from unfertilized and fertilized sea urchin eggs and its effect on maternal m-RNA translation. Biochim. biophys. Acta (Amst.) 312, 377-391 (1973).

Gartler, S.M., Liskay, R.M., Campbell, B.K., Sparkes, R., Gant, N.: Evidence for two functional X-chromosomes in human oocytes. Cell Differ. 1, 215-218 (1972).

Gartler, S.M., Liskay, R.M., Gant, N.: Two functional X-chromosomes in human fetal oocytes. Exp. Cell Res. 82, 464-466 (1973).

Gates, A.H.: Maximizing yield and developmental uniformity of eggs. In: Methods in Mammalian embryology. San Francisco: Freeman and Co. 1971.

Glass, L.E.: Transmission of maternal proteins into oocytes. Advanc. Biosci. 6, 29-61 (1971).

Glisin, V.R., Glisin, M.V.: Ribonucleic acid and metabolism following fertilization in sea urchin eggs. Proc. nat. Acad. Sci. (Wash.) 52, 1548-1553 (1964).

Golbus, M.S., Calarco, P.G., Epstein, C.J.: The effects of inhibitors of RNA synthesis (α-amanitin and actinomycin D) on preimplantation mouse embryogenesis. J. exp. Zool. 186, 207-216 (1973).

Graham, C.F.: Parthenogenetic mouse blastocysts. Nature (Lond.) 226, 165-167 (1970).

Graham, C.F.: The production of parthenogenetic mammalian embryos and their use in biological research. Biol. Rev. 49, 399-422 (1974).

Graham, C.F., Deussen, Z.A.: In vitro activation of mouse eggs. J. Embryol. exp. Morph. 31, 497-512 (1974).

Greenberg, J.R., Perry, R.P.: Relative occurrence of polyadenylic acid sequences in messenger and heterogeneous nuclear RNA of L-cells as determined by poly (U) hydroxylapatite chromatographie. J. molec. Biol. 72, 91-98 (1972).

Greenwald, G.S., Everett, N.B.: The incorporation of S^{35} methionine by the uterus and ova of the mouse. Anat. Rec. 134, 171-184 (1959).

Gropp, A.: Reproductive failure due to fetal aneuploidy in mice. In: Fertility and Sterility, Proceedings of the VIII World Congress. Amsterdam: Excerpta Medica 1973.

Gross, K.W., Jacobs-Lorena, J., Baglioni, C., Gross, P.R.: Cell-free translation of maternal messenger RNA from sea urchin egg. Proc. nat. Acad. Sci. (Wash.) 70, 2614-2618 (1973b).

Gross, P.R.: The control of protein synthesis in embryonic development and differentiation. Curr. Top. develop. Biol. 2, 1-43 (1967a).

Gross, P.R.: RNA metabolism in embryonic development and differentiation. II. Biosynthetic patterns and their regulation. New Engl.J. Med. 276, 1297-1305 (1967b).

Gross, P.R.: Biochemistry of differentiation. Ann. Rev. Biochem. 37, 631-660 (1968).

Gross, P.R.: Protein synthesis during cleavage. In: Molecular Genetics and Developmental Biology. Englewood Cliffs: Prentice-Hall, Inc. 1972.

Gross, P.R., Cousineau, G.: Effects of actinomycin D on macromolecule synthesis and early development in sea urchin eggs. Biochem. biophys. Res. Commun. 10, 321-326 (1963).

Gross, P.R., Cousineau, G.: Macromolecule synthesis and the influence of actinomycin D on early development. Exp. Cell Res. 33, 368-395 (1964).

Gross, P.R., Fry, B.J.: Continuity of protein synthesis through cleavage metaphase. Science 153, 749-751 (1966).

Gross, P.R., Gross, K.W., Skoultchi, A.I., Ruderman, J.V.: Maternal mRNA and protein synthesis in the embryo. In: Protein Synthesis in Reproductive Tissue. Stockholm: Karolinska Institutet 1973.

Gurley, L.R., Walters, R.A., Tobey, R.A.: Histone phosphorylation in late interphase and mitosis. Biochem. biophys. Res. Commun. 50, 744-750 (1973).

Gwatkin, R.B.L., Williams, D.T.: Heat sensitivity of the cortical granule protease from hamster eggs. J. Reprod. Fertil. 39, 153-155 (1974).

Gwatkin, R.B., Williams, D.T., Hartmann, J.F., Kniazuk, M.: The zona reaction of hamster and mouse eggs: Production in vitro by a trypsin-like protease from cortical granules. J. Reprod. Fertil. 32, 259-265 (1973).

Halberg, R.L., Smith, D.C.: Ribosomal protein synthesis in Xenopus laevis oocytes. Develop. Biol. 42, 40-52 (1975).

Hansmann, I., Röhrborn, G.: Chromosome aberrations in preimplantation stages of mice after treatment with Triazoquinone. Humangenetik 18, 101-109 (1973).

Hedberg, E.: The chemical composition of the human ovarian oocyte. Acta endocr. (Kbh.) 14 (Suppl. 15) 1-89 (1953).

Hennen, S.: Chromosomal and embryological analyses of nuclear changes occurring in embryos derived from transfers of nuclei between Rana pipiens and Rana sylvatica. Develop. Biol. 6, 133-183 (1963).

Hille, M.B.: Inhibitor of protein synthesis isolated from ribosomes of unfertilised eggs and embryos of sea urchins. Nature (Lond.) 249, 556-558 (1974).

Hitzeroth, H., Klose, J., Ohno, S., Wolf, U.: Asynchronous activation of parental alleles at the tissue specific gene loci observed on hybrid trouts during early embryonic development. Biochem. Genet. 1, 287-300 (1968).

Hogan, B., Gross, P.R.: The effect of protein synthesis inhibition on the entry of messenger-RNA into the cytoplasm of sea urchin embryos. J. Cell Biol. 49, 692-701 (1971).

Honig, G.R., Rabinowitz, M.: Actinomycin D inhibition of protein synthesis unrelated to effect on template RNA synthesis. Science 149, 1504-1506 (1965).

Hultin, T.: The effect of puromycin on protein metabolism and cell division in fertilized sea urchin eggs. Experientia (Basel) 17, 410-411 (1961).

Humphreys, T.: Efficiency of translation of messenger RNA before and after fertilization on sea urchins. Develop. Biol. 20, 435-458 (1969).

Humphreys, T.: Mesurements of messenger RNA entering polysomes upon fertilization of sea urchin. Develop. Biol. 26, 201-208 (1971).

Hynes, R.O., Gross, P.R.: Informational RNA sequences in early sea urchin embryos. Biochim. biophys. Acta (Amst.) 259, 104-111 (1972).

Jacob, S.E., Sajdel, W., Muecke, W., Munro, H.: Soluble RNA polymerases of rat liver nuclei: Properties, template specificity, and α-amanitin responses in vitro and in vivo. Cold Spr. Harbor Symp. quant. Biol. 35, 681-691 (1970).

Johnson, A.W., Hnilica, L.S.: Cytoplasmic and nuclear basic protein synthesis during early sea urchin development. Biochim. biophys. Acta (Amst.) 246, 141-154 (1971).

Kedes, L., Gross, P.R.: Synthesis and function of messenger RNA during early embryonic development. J. molec. Biol. 42, 559-575 (1969).

de Kerckhove, D. van: Content of deoxyribonucleic acid of the germinal vesicle of the primary oocyte in the rabbit. Nature (Lond.) 183, 329 (1959).

Kuwahara, M., Chaykin, S.: Biosynthesis of pyridine nucleotides in early embryos of the mouse (Mus musculus). J. biol. Chem. 248, 5095-5099 (1973).

LaMarca, M.J., Smith, L.D., Strobel, M.C.: Quantitative and qualitative analysis of RNA synthesis in stage 6 and stage 4 oocytes of Xenopus laevis. Develop. Biol. 34, 106-118 (1973).

Laszlo, J., Miller, D.S., McCarty, K.S., Hochstein, P.: Actinomycin D: Inhibition of respiration and glycolysis. Science 151, 1007-1010 (1966).

Lee, S.Y., Mendecki, J., Brawerman, G.: A polynucleotide segment rich in adenylic acid in the rapidly labeled polyribosomal RNA component of mouse sarcoma 180 ascites cells. Proc. nat. Acad. Sci. (Wash.) 68, 1331-1335 (1971).

Lewin, B.: Gene expression – 2. Eucaryotic chromosomes. London-New York-Sydney-Toronto: Wiley and Sons 1974.

Lim, L., Canellakis, E.S.: Adenine-rich polymer associated with rabbit reticulocytes messenger RNA. Nature (Lond.) 227, 710-712 (1970).

Lin, T.P.: DL-methionine (sulphur $^{-35}$) for labelling unfertilized mouse eggs in transplantation. Nature (Lond.) 178, 1175-1176 (1956).

Lindell, T.J., Weinberg, F., Morris, P.W., Roeder, R.G., Rutter, W.J.: Specific inhibition of nuclear RNA polymerase II by α-amanitin. Science 170, 447-449 (1970).

Lintern-Moore, S., Peters, H., Moore, G.P.M., Faber, M.: Follicular development in the infant human ovary. J. Reprod. Fertil. 39, 53-64 (1974).

Loewenstein, J.E., Cohen, A.I.: Dry mass, lipid content and protein content of the intact and zona-free mouse ovum. J. Embryol. exp. Morph. 12, 113-121 (1964).

Lutwak-Mann, C.: Some physiological and biochemical properties of the mammalian blastocyst. Bull. schweiz. Akad. med. Wiss. 22, 101-122 (1966).

MacGregor, H.C.: The role of lampbrush chromosomes in the formation of nucleoli in amphibian oocytes. Quart. J. micr. Sci. 106, 215-228 (1965).

MacGregor, H.C.: Nucleolar DNA in oocytes of Xenopus laevis. J. Cell Sci. 3, 437-444 (1968).

Maggio, R., Catalano, C.: Activation of amino-acids during sea urchin development. Arch. Biochem. Biophys. 103, 164-168 (1963).

Mairy, M., Denis, H.: Recherches biochimiques sur l'oogenèse. I. Synthese et accumulation du RNA pendant l'oogenèse du crapaud sudafricain Xenopus laevis. Develop. Biol. 24, 143-165 (1971).

Manes, C.: Nucleic acid synthesis in preimplantation rabbit embryos. I. Quantitative aspects, relationship to early morphogenesis and protein synthesis. J. exp. Zool. 172, 303-310 (1969).

Manes, C.: Nucleic acid synthesis in preimplantation rabbit embryos. II. Delayed synthesis of ribosomal RNA. J. exp. Zool. 176, 87-95 (1971).

Manes, C.: The participation of the embryonic genome during early cleavage in the rabbit. Develop. Biol. 32, 453-459 (1973).

Manes, C., Daniel, J.C., Jr.: Quantitative and qualitative aspects of protein synthesis in the preimplantation rabbit embryo. Exp. Cell Res. 55, 261-268 (1969).

Mangia, F., Epstein, C.J.: Biochemical studies of growing mouse oocytes: Preparation of oocytes and analysis of glucose-6-phosphate dehydrogenase and lactate dehydrogenase activities. Develop. Biol. 45, 211-220 (1975).

Mescher, A., Humphreys, T.: Activation of maternal mRNA in the absence of poly (A) formation in fertilized sea urchin eggs. Nature (Lond.) 249, 138-139 (1974).

Metafora, S., Felicetti, L., Gambino, R.: The mechanism of protein synthesis activation after fertilization of sea urchin eggs. Proc. nat. Acad. Sci. (Wash.) 68, 600-604 (1971).

Milcarek, C., Price, P., Penman, S.: The metabolism of a poly (A) minus fraction in HeLa cells. Cell 3, 1-10 (1974).

Mintz, B.: Synthetic processes and early development in the mammalian egg. J. exp. Zool. 157, 85-100 (1964).

Monesi, V., Molinaro, M., Spalletta, E., Davoli, C.: Effect of metabolic inhibitors on macromolecular synthesis and early development in the mouse embryo. Exp. Cell Res. 59, 197-206 (1970).

Monroy, A., Maggio, R., Rinaldi, A.M.: Experimentally induced activation of the ribosomes of the unfertilized sea urchin egg. Proc. nat. Acad. Sci. (Wash.) 54, 107-111 (1965).

Monroy, A., Vittorelli, M.: Utilization of C^{14}-glucose for amino acids and protein synthesis by the sea urchin embryo. J. cell. comp. Physiol. 60, 285-287 (1962).

Moore, G.P.M., Lintern-Moore, S.: A correlation between growth and RNA synthesis with mouse oocyte. J. Reprod. Fertil. 39, 163-166 (1974).

Moore, G.P.M., Lintern-Moore, S., Peters, H., Faber, M.: RNA synthesis in the mouse oocyte. J. cell Biol. 60, 416-422 (1974).

Moore, R.W., Brinster, R.L.: Transamination and deamination in mouse preimplantation embryos. Proc. Soc. Study Reprod. Columbus, Ohio (1970).

Morris, T.: The XO and YO chromosome constitutions in the mouse. Genet. Res. 12, 125-137 (1968).

Noronha, J.M., Sheys, G.H., Buchanan, J.M.: Induction of a reductive pathway for deoxyribonucleotide synthesis during early embryogenesis of the sea urchin. Proc. nat. Acad. Sci. (Wash.) 69, 2006-2010 (1972).

Oakberg, E.F.: ^{3}H-uridine labeling of mouse oocytes. Arch. Anat. micr. Morph. exp. 56 (Suppl. 3-4), 171-184 (1967).

Oakberg, E.F.: Relationship between stage of follicle development and RNA synthesis in the mouse oocyte. Mutation Res. 6, 155-165 (1968).

Odeblad, E., Magnusson, G.: An autoradiographic study on the intracellular accumulation of radioactive phosphate in the egg cell of the mouse. Acta endocr. (Kbh.) 17, 290-293 (1954).

Ohno, S.: Sex chromosomes and sex-linked genes. Berlin-Heidelberg-New York: Springer 1967.

Ohno, S.; The preferential activation of maternally derived alleles in development of interspecific hybrids. In: Heterospecific genome interaction. Philadelphia: The Wistar Institute Press 1969.

Ohno, S.: Ancient linkage groups and frozen accidents. Nature (Lond.) 244, 259-262 (1973).

Olds, P.J., Stern, S., Biggers, J.D.: Chemical estimates of the RNA and DNA contents of the early mouse embryo. J. exp. Zool. 186, 39-46 (1973).

Panje, W.R., Kessel, R.G.: Soluble proteins and quantitative analysis of protein and RNA during oogenesis in Necturus maculosus. Exp. Cell Res. 51, 313-322 (1968).

Pastan, I., Friedman, R.M.: Actinomycin D: Inhibition of phospholipid synthesis in chick embryo cells. Science 160, 316-317 (1968).

Pedersen, T., Peters, H.: Proposal for a classification of oocytes and follicles in the mouse ovary. J. Reprod. Fertil. 17, 555-557 (1968).

Perkowska, E., MacGregor, H.C., Birnstiel, M.L.: Gene amplification in the oocyte nucleus of mutant and wild-type Xenopus laevis. Nature (Lond.) 217, 649-650 (1968).

Petzoldt, U., Dames, W., Gottschewski, G.H.M., Neuhoff, V.: Das Proteinmuster in frühen Entwicklungsstadien des Kaninchens. Cytobiologie 5, 272-280 (1972).

Piatigorsky, J.: Ribonuclease and trypsin treatment of ribosomes and polyribosomes of sea urchin eggs. Biochim. biophys. Acta (Amst.) 166, 142-155 (1968).

Piatigorsky, J., Tyler, A.: Changes upon fertilization in the distribution of RNA containing particles in sea urchin eggs. Develop. Biol. 21, 13-28 (1970).

Poznakhirkina, N.A., Serov, O.L., Korochkin, L.I.: A study on lactate dehydrogenase isozymes in rat ova. Biochem. Genet. 13, 65-72 (1975).

Quinn, P., Kozak, L.P.: Activity of certain glycolytic enzymes in developing mouse embryos. J. Reprod. Fertil. 43, 388-389 (1975).

Quinn, P., Wales, R.G.: Fixation of carbon dioxide by preimplantation mouse embryos in vitro and the activities of enzymes involved in the process. Aust. J. biol. Sci. 24, 1277-1290 (1971).

Raff, R.A., Colot, H.U., Selwig, S.E., Gross, P.R.: Oogenetic origin of messenger RNA for embryonic synthesis of microtubule proteins. Nature (Lond.) 235, 211-214 (1972).

Raff, R.A., Greenhouse, G.A., Gross, K.W., Gross, P.R.: Synthesis and storage of microtubule proteins by sea urchin embryos. J. Cell Biol. 50, 516-527 (1971).

Rapola, J., Koskimies, O.: Embryonic enzyme patterns: Characterization of the single lactate dehydrogenase isozyme in preimplanted mouse ova. Science 157, 1311-1312 (1967).

Reamer, G.R.: The quantity and distribution of nucleic acids in the early cleavage stages of the mouse embryo. Ph.D. Thesis. Boston University 1963.

Revel, M., Hiatt, H.H., Revel, J.P.: Actinomycin D: An effect on rat liver homogenates unrelated to its action on RNA synthesis. Science 146, 1311-1313 (1964).

Rinaldi, A.M., Monroy, A.: Polyribosome formation and RNA synthesis in early post-fertilization stages of sea urchin egg. Develop. Biol. 19, 73-86 (1969).

Roeder, R.G., Chou, S., Jaehning, J.A., Schwarz, L.B., Sklar, V.E.F., Weinmann, R.: Structure, function and regulation of RNA polymerases in animal cells. In: Isozymes III: Developmental Biology. New York: Academic Press 1975.

Rosbash, M., Ford, P.J.: Polyadenylic acid containing RNA in Xenopus laevis oocytes. J. molec. Biol. 85, 87-101 (1974).

Roversi, G.D., Silvestrini, R.: Study on the protein metabolism of the evolutional ovarian follicle. Autoradiographic research with ^{3}H-phenylalanine. Exp. Cell Res. 31, 484-489 (1963).

Ruderman, J.V., Gross, P.R.: Histones and histone synthesis in sea urchin development. Develop. Biol. 36, 286-298 (1974).

Samarina, O.P., Lukanidin, E.M., Georgiev, G.P.: Ribonucleoprotein particles containing mRNA and pre-mRNA. In: Protein synthesis in reproductive tissue. Stockholm: Karolinska Institutet 1973.

Scherrer, K.: Messenger RNA in eukaryotic cells: The life history of duck globin messenger RNA. In: Protein Synthesis in Reproductive Tissue. Stockholm: Karolinska Institutet 1973.

Schlafke, S., Enders, A.C.: Cytological changes during cleavage and blastocyst formation in the rat. J. Anat. 102, 13-32 (1967).

Schuel, H., Wilson, W.L., Chen, K., Lorand, L.: A trypsin-like proteinase localized in cortical granules isolated from unfertilized sea urchin eggs by zonal centrifugation. Role of the enzyme in fertilization. Develop. Biol. 34, 175-186 (1973).

Schultz, G.A.: Characterization of polyribosomes containing newly-synthesized messenger RNA in preimplantation rabbit embryos. Exp. Cell Res. 82, 168-174 (1973).

Schultz, G.A.: The stability of messenger RNA containing polyadenylic acid sequences in rabbit blastocysts. Exp. Cell Res. 86, 190-193 (1974).

Schultz, G.A.: Polyadenylic acid-containing RNA in unfertilized and fertilized eggs of the rabbit. Develop. Biol. 44, 270-277 (1975).

Schultz, G.A., Manes, C., Hahn, W.E.: Synthesis of RNA containing polyadenylic acid sequences in preimplantation rabbit embryos. Develop. Biol. 30, 418-426 (1973).

Sheldon, R., Jurale, C., Kates, J.: Detection of polyadenylic acid sequences in viral and eucaryotic RNA. Proc. nat. Acad. Sci. (Wash.) 69, 417-421 (1972).

Siracusa, G.: RNA polymerase during early development in mouse embryo. Exp. Cell Res. 78, 460-462 (1973).

Siracusa, G., Vivarelli, E.: Low-salt and high-salt RNA polymerase activity during preimplantation development in the mouse. J. Reprod. Fertil. 43, 567-569 (1975).

Sirlin, J.L.: Biology of RNA. New York–London: Academic Press 1972.

Skalko, R.G., Morse, J.M.: The differential response of the early mouse embryo to Actinomycin D treatment in vitro. Teratology 2, 47-54 (1969).

Skoultchi, A., Gross, P.R.: Maternal histone messenger RNA: detection by molecular hybridization. Proc. nat. Acad. Sci. (Wash.) 70, 2840-2844 (1973).

Slater, I., Gillespie, D., Slater, D.W.: Cytoplasmic adenylation and processing of maternal RNA. Proc. nat. Acad. Sci. (Wash.) 70, 406-411 (1973).

Slater, I., Slater, D.W.: Polyadenylation and transcription following fertilization. Proc. nat. Acad. Sci. (Wash.) 71, 1103-1107 (1974).

Smith, L.D.: Protein synthesis during oocyte maturation. In: Oogensis. Baltimore: University Park Press 1972.

Smith, L.D., Ecker, R.E.: Protein synthesis in enucleated eggs of Rana pipiens. Science 150, 777-779 (1965).

Soeiro, R., Amos, H.: mRNA half-life measured by the use of Actinomycin D in animal cells–a caution. Biochim. biophys. Acta (Amst.) 129, 406-409 (1966).

Sorensen, R.A.: Problems in oocyte maturation and early development in the mouse. Ph.D. Thesis, Yale University 1972.

Spielmann, H.: Different patterns of energy metabolism in the rat and mouse zygote. J. Reprod. Fertil. 42, 391-394 (1975).

Spielmann, H., Erickson, R.P., Epstein, C.J.: Immunological studies of lactate dehydrogenase and glucose-6-phosphate dehydrogenase in preimplantation mouse embryo. J. Reprod. Fertil. 40, 367-373 (1974).

Spirin, A.S.: Informosomes. Europ. J. Biochem. 10, 20-35 (1969).

Spirin, A.S.: Nonribosomal RNP particles (informosomes) of animal cells. In: The Mechanism of Protein Synthesis and its Regulation. Amsterdam: North Holland 1972.

Stahl, A., Luciani, J.M., Devictor, M., Capordano, A.M., Gagné, R.: Constitutive heterochromatin and micronucleoli in the human oocyte at the diplotene stage. Humangenetik 26, 315-327 (1975).

Stambaugh, R., Buckley, B.S.: Histochemical subcellular localization of the acrosomal proteinase effecting dissolution of the zona pellucida using fluorescein-labeled inhibitors. Fertil.and Steril. 23, 348-352 (1972).

Steinberg, R.A., Levinson, B.B., Tomkins, G.M.: "Superinduction" of tyrosine aminotransferase by actinomycin D: a reevaluation. Cell 5, 29-39 (1975).

Stern, S.: The activity of glycogen synthetase in the cleaving mouse embryo. Proc. Soc. Study Reprod. Columbus, Ohio (1970).

Stern, S., Raygis, A., Kennedy, J.F.: Incorporation of amino acids during maturation in vitro by mouse oocytes: effect of puromycin on protein synthesis. Biol. Reprod. 7, 341-346 (1972).

Streffer, C., Beuningen, D. van: The content of pyridine dinucleotides in the mouse embryo before implantation. Develop. Biol. 38, 401-403 (1974).

Tarkowski, A.K., Witkowska, A., Nowicka, J.: Experimental parthenogenesis in the mouse. Nature (Lond.) 226, 162-165 (1970).

Tasca, R.J., Hillman, N.: Effects of Actinomycin D and cykloheximide on RNA and protein synthesis in cleavage stage mouse embryos. Nature (Lond.) 225, 1022-1025 (1970).

Terman, S., Gross, P.: Translation -- level control of protein synthesis during early development. Biochem. biophys. Res. Commun. 21, 595-600 (1966).

Thomson, J.L., Biggers, J.D.: Effect of inhibitors of protein synthesis on the development of preimplantation mouse embryos. Exp. Cell Res. 41, 411-427 (1966).

Tyler, A.: Masked messenger RNA and cytoplasmic DNA in relation to protein synthesis and processes of fertilization and determination in embryonic development. Develop. Biol., Suppl. 1, 170-226 (1967).

Vacquier, V.D., Epel, D., Douglas, L.: Sea urchin eggs release protease activity at fertilization. Nature (Lond.) 237, 34-36 (1972).

Wallace, R.A., Jared, D.W.: Studies on amphibian yolk. VIII. The estrogen – induced hepatic synthesis of a serum lipophosphoprotein and its selective uptake by the ovary and transformation into yolk platelet protein in Xenopus laevis. Develop. Biol. 19, 498-526 (1969).

Weakley, B.S.: Electron microscopy of the oocyte and granulosa cells in the developing ovarian follicle of the Golden hamster (Mesocricetus auratus). J. Anat. 100, 503-534 (1966).

Weakley, B.S.: Comparison of cytoplasmatic and membraneous elements in the oocyte of five mammalian species. Z. Zellforsch. 85, 109-123 (1968).

Whiteley, A.H., McCarthy, B.J., Whiteley, H.R.: Changing populations of messenger RNA during sea urchin development. Proc. nat. Acad. Sci. (Wash.) 55, 519-525 (1966).

Whitt, G.S., Childers, W.F., Cho, P.L.: Allelic expression at enzyme loci in an intertribal hybrid sunfish. J. Hered. 64, 55-61 (1973).

Whitten, W.K.: Culture of tubal ova. Nature (Lond.) 179, 1081-1082 (1957).

Whittingham, D.G.: The failure of lactate and phosphoenolpyruvate to support development of the mouse zygote in vitro. Biol. Reprod. 1, 381-386 (1969).

Wilson, E.B.: On cleavage and mosaic work. Arch. Entwickl.-Mech. Org. 3, 19-30 (1896).

Wilt, F.H.: Polyadenylation of maternal RNA of sea urchin eggs after fertilization. Proc. nat. Acad. Sci. (Wash.) 70, 2345-2349 (1973).

Wolf, U., Engel, W.: Gene activation during early development of mammals. Humangenetik 15, 99-118 (1972).

Young, R.J., Stull, G.B., Brinster, R.L.: RNA in mouse ovulated oocytes. J. Cell Biol. 59, 372a (1973).

Zamboni, L., Mastroianni, L.: Electron microscopic studies on rabbit ova. J. Ultrastruct. Res. 14, 95-117 (1966).

Zeilmaker, G.H., Hulsman, W.C., Wensinck, F., Verhamme, C.: Oxygen-triggered mouse oocyte maturation in vitro and lactate utilization by mouse oocytes and zygotes. J. Reprod. Fertil. 29, 151-152 (1972).

The Genetic Activity of Early Mammalian Embryos

CH. J. EPSTEIN

What is the evidence that gene expression has anything to do with early mammalian development – specifically with preimplantation development? Much is already known about biochemical and metabolic changes which occur during the preimplantation period, and in this discussion I shall outline some of this work and view it from the perspective of what it tells us about gene activity (for a more comprehensive review, see *Epstein, C.J.*, 1975).

Many enzymes are detectable in the preimplantation mouse embryo, and several undergo significant changes in their activities. Some, such as lactate dehydrogenase, glucose-6-phosphate dehydrogenase, and guanine deaminase, decrease in activity during the latter part of the preimplantation period, while others, such as malate dehydrogenase and aldolase, stay relatively constant. Several enzymes – for example, hypoxanthine guanine phosphoribosyltransferase, adenyl phosphoribosyltransferase, and uridine kinase – increase in activity. There is no single consistent pattern of enzyme change or development during the preimplantation period, at least in so far as activity is concerned, and the different enzymes behave in different ways (*Epstein* and *Daentl,* 1972).

Among the other biochemical systems that have been studied during the preimplantation period of the mouse are those which transport macromolecular precursors and energy sources. It is not as yet known whether these transport systems are actually operating in vivo and whether they are necessary during this period of development. Nonetheless, the embryo, as it develops, is changing significantly with regard to its capabilities for transport, and we assume that these changes are determined by molecular changes in the membrane. Transport capacities for several RNA precursors (nucleotide bases and nucleosides) have been examined, and these substances appear to fall into three different groups distinguishable by developmental and competition studies (*Epstein* and *Daentl*, 1972). One is the group of the nucleosides, for which there is very little capacity for the transport between the two cell and the eight cell stages (a fact which has bedeviled the work on nucleic acid synthesis during the very early stages of embryonic development). There is, however, a great increase in the capacity for transport of all nucleosides between the 8-16 cell and the early blastocyst stages and then, as the blastocyst expands, transport capacity levels off or actually declines. A second group is represented by the purine bases, adenine and guanine, and reaches at least two-thirds or three-fourths of its maximum activity between the two-cell and the eight-cell stages. Between the eight cell and early blastocyst stages it remains level or increases slightly and then falls off quite sharply. The third system, that of the pyrimidine bases, is extremely low in capacity and increases only slightly during the early embryonic period. Again, as in the case of the enzymes, there is no "master switch" for transport: each of these transport systems, as well as others which have been analyzed, changes in its own way and at its own time (*Kramen* and *Biggers*, 1971; *Epstein* and *Smith*, 1973).

Unfortunately, none of this information about enzymes or transport systems proves that genes have been activated or are even expressing themselves during the preimplantation period. All of these changes could take place through the activation or inactivation of molecules or systems which already exist in the embryo. And, even if the observed changes did result from the synthesis of new proteins, this might still not be determined by activity of the genetic apparatus of the embryo per se and could result from activation of genetic information already laid down in the egg. In lower organisms (sea urchins, amphibians) there is considerable evidence for the existence of forms of messenger RNA, the so-called maternal masked messenger RNAs, which are present in the egg and are activated during the early period of postfertilization (*Gross*, 1967). Nothing discussed to this point argues for or against the activity of such masked messengers in the early developing mouse embryo (see also *Engel* and *Franke*, this volume).

Going beyond these types of investigations, which are essentially studies of the most distant biochemical representations of embryonic genetic activity, it is possible to look at processes which are closer to gene expression. During the early period of development, from the ovulated egg to about the eight cell stage, there is a low but easily measured incorporation of amino acids into protein, with perhaps a small increase between the two-cell and eight-cell stages. Then, in the period between the eight-cell and early blastocyst stages, there is a large increase in the amount of amino acid incorporated into protein which parallels the increase in amino acid uptake of the precursor. However, the two processes become dissociated from one another as the blastocyst grows, so that the uptake of precursor, a measure of transport, continues to increase while the incorporation into protein levels off. By appropriate consideration of the state of the precursor within the embryo, it has been shown that these changes in the rate of incorporation of amino acid into protein represent real changes in the rate of protein synthesis and are not merely representations of changes in the rate of uptake (*Epstein* and *Smith*, 1973). However, the proteins which are actually being synthesized are not known. Although there are enzymes which are increased in activity during this period, they could represent only a small fraction of the protein that is synthesized. The real difficulty in knowing what proteins are being made is in deciding which proteins to look at. Since it is not easy to predict in advance what proteins might be made, more general approaches have been used to look at a wide variety simultaneously. One such technique is that of protein fractionation on polyacrylamide gels containing sodium dodecyl sulfate. This system solubilizes virtually all proteins and fractionates them according to their molecular sizes or weights. If mouse embryos are cultured in the presence of ^{3}H- or ^{14}C-labeled amino acids, specific electrophoretic patterns of incorporation of amino acids into protein are obtained (*Epstein* and *Smith*, 1974). This incorporation represents true protein synthesis and can be blocked by protein synthesis inhibitors such as cycloheximide. Several peaks which have not as yet been identified can be observed over the molecular weight range between 30,000 and 150,000 daltons, and the pattern in the early blastocyst is quite different from that in two-cell embryos. To facilitate comparisons, a double labeling system has been used in which one stage is labeled with ^{3}H-amino acid and another with ^{14}C amino acid. Embryos of the two stages are then mixed and run through the analytical procedure simultaneously. By this method, there appears to be an increase in synthesis of certain components with molecular weights of about 35,000 or 40,000 and of 68,000 daltons and a large decrease in the synthesis of a 55,000 dalton component (Fig. 1).

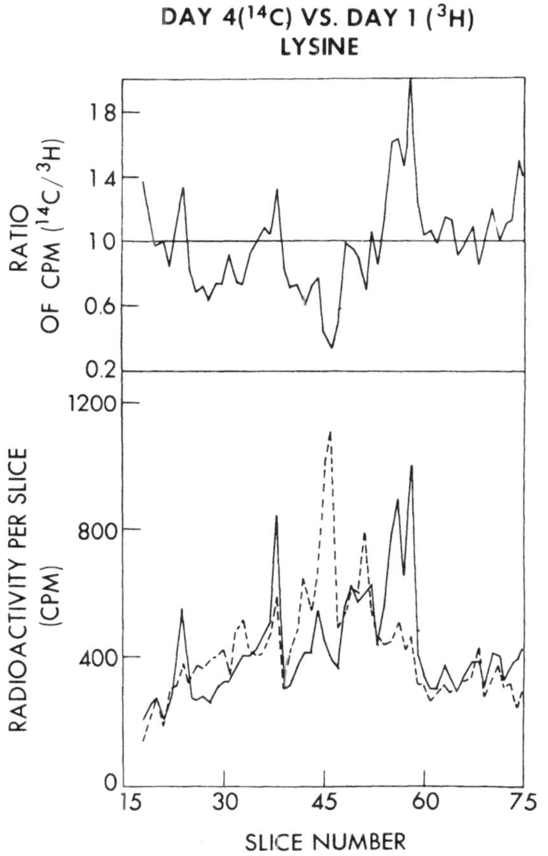

Fig. 1. Incorporation of lysine into proteins of day 1 (2 cell) and day 4(late blastocysts) mouse embryos. Day 1 embryos incubated in ^{3}H-lysine and day 4 in ^{14}C-lysine were combined, solubilized in sodium dodecyl sulfate, and electrophoresed in SDS-polyacrylamide gel. Gel was then sliced and each slice analyzed for ^{3}H and ^{14}C radioactivity. Top: Normalized ratio of ^{14}C (day 4) to ^{3}H (day 1) radioactivity, indicating significant changes in relative rates of synthesis of many protein constituents. Bottom: Radioactivity of each gel slice; solid line represents ^{14}C (day 4) protein and dashed line ^{3}H (day 1) protein. Major day 4 peaks have apparent molecular weights of 68,000 (slice 39), 39,000 (slice 56), and 36,000 (slice 58) daltons; major day 1 peak is 55,000 daltons (slice 47) (*Epstein* and *Smith*, 1974)

Thus, not only do enzymes and transport systems change during the preimplantation period, but so also does the synthesis of the proteins of the embryos. But, by itself this again does not constitute direct evidence for the role of gene activity, since these changes in protein synthesis could also be determined by differences in the translation of pre-existing messenger RNAs. Therefore, to get closer to the gene, it is necessary to examine the synthesis of RNA itself. As with protein synthesis, there is a large increase between the eight-cell stage and the blastocyst stages in the total amount of RNA that is being made (*Epstein* and *Daentl*, 1971). With careful labeling techniques, however, it has been possible to show that RNA synthesis is demonstrable even as early as the two-cell stage. By the four-cell stage, the various common types of RNA (transfer RNA, ribosomal RNA, high molecular weight RNA) are being made in the mouse embryo, and these continue to be made, in changing proportions, during the entire preimplantation period (*Ellem* and *Gwatkin*, 1968; *Woodland* and *Graham*, 1969). It is clear, therefore, that RNA synthesis does change and increase during the preimplantation period, but again there is the question of the significance of these changes. Do these new RNAs really have a determinative role in the developmental process? To get at this type of information more directly, use has been made of inhibitors of RNA synthesis and of gene expression. Several studies have been carried out with actinomycin D, and it has been shown that this drug inhibits the cleavage of mouse embryos at whatever stage given, even as early as the one-cell and two-cell stages (*Golbus* and *Epstein*, 1973). Unfortunately, work with actinomycin D has been criticized because it has effects which go beyond inhibition of RNA synthesis. Recently, a new RNA synthesis inhibitor has appeared. This drug, α-amanitin, is a specific in vitro inhibitor of type II RNA polymerase, the polymerase considered responsible for messenger RNA synthesis. At reasonably low concentrations, α-amanitin inhibits the development of preimplantation mouse embryos. Thus, when embryos between the 4- and 12-cell stage are cultured in α-amanitin, they develop to the morula stage, which represents one or at the most two cleavages, but blastulation is severely inhibited (*Golbus* and *Epstein*, 1973). These findings suggest that messenger RNA synthesis and expression are necessary for blastocyst formation.

Another drug that has been used in several systems is bromodeoxyuridine (BrdU). The mechanism of BrdU action is not well understood, although incorporation into DNA appears to be necessary. Two cell mouse embryos cultured in BrdU develop as far as the morula stage but will not blastulate. The same also occurs with four- to eight-cell embryos, and even the blastulation of embryos started in culture as morulae is reduced (*Golbus* and *Epstein*, 1974). Reasoning from what is known about other developmental systems, it may be inferred that some genetic activity which BrdU is affecting is required for the formation of what is morphologically interpreted as a blastocyst.

With these inhibitor studies it is possible to come closer to the conclusion that genetic activity is required for early development in the mouse embryo. There is, however, even more direct evidence. First, there are mutations which affect development during the early preimplantation period. The best known of these is the t^{12} mutation, which in the homozygous form prevents development beyond the morula stage. It has recently been shown that a gene at the *T*-locus is associated with a surface antigen that makes its appearance in the embryo prior to and has its maximum concentration at the eight-cell stage, just before the morula is formed (*Artzt* et al., 1973). This work indicates that there is a preimplantation embryo surface constituent which is synthesized under genetic control and that synthesis of abnormal material is associated with a specific defect in development.

A second direct approach to embryonic genetic activity is represented by the investigations that have been carried out to demonstrate the activation of the *paternal* genome during the preimplantation period. Since the paternal genome is not present in the egg, its expression in the embryo must represent activation of the embryonic genome. Two systems have been looked at. One is the enzyme glucose phosphate isomerase, the paternal form of which appears at the late morula or early blastocyst stage (*Brinster*, 1973). The other, β-glucuronidase, has been shown to appear in the paternal form by about the eight-cell stage (*Chapman* and *Wudl*, 1974). From studies of these two enzymes, therefore, it is clear that activation of the embryonic genome has occurred quite soon after fertilization.

Another approach has made use of a more indirect type of analysis. When eggs are obtained from female mice which are either XO or XX in constitution, both fertile but differing in the number of X chromosomes present, it is found that the activity of hypoxanthine guanine phosphoribosyltransferase (HGPRT) is half as great in the eggs of XO mice as in the eggs of XX females. Taken in conjuction with the known X-linkage of HGPRT in man, this evidence indicates that both mouse X-chromosomes function during normal oogenesis (*Epstein*, 1972; see also *Engel* and *Franke*, this volume). Embryonic HGPRT increases in activity during the preimplantation period, with a six to eight-fold rise occurring between the two cell and early blastocyst stages (*Epstein*, 1970). If messenger RNA for the synthesis of this enzyme is made in the *egg* and is being read out later, the same dosage relationship as was found in eggs of XX and XO females would be expected to hold. On the other hand, if the messenger is made after fertilization, then, even taking into account the mixed progeny of the XO females, the activities of HGPRT in embryos of XO and XX females should be close to one another. When examined directly, no substantial difference was found between the activity of HGPRT in the blastocysts of XO and XX mothers, whereas the ratio is about 1:2 in ovulated eggs (*Epstein*, 1972). This again constitutes evidence that new postfertilization messengers are being synthesized during the period preceding the formation of the blastocyst.

Before concluding, brief reference to the rabbit is in order. Considerable work has been carried out with the developing rabbit embryo, which, because of its size is susceptible to more sophisticated types of biochemical analysis than is the mouse embryo. Although synthesis of some forms of RNA, such as ribosomal RNA, is considerably delayed in the rabbit embryo as it is compared with the mouse, α-amanitin inhibits cleavage in a similar manner (*Manes*, 1973). The existence of RNA containing poly A, which appears to be a marker for messenger RNA, can be demonstrated as early as the 16 cell stage in the rabbit embryo (*Schultz* et al., 1973a). And, although the preimplantation period per se has not been examined, comparison of the late blastocyst stage with the day 12 rabbit embryo demonstrates differences in the sequences of DNA that are actually represented in the RNA (*Schultz* et al., 1973b). Thus, the evidence in the rabbit, coupled with that already described in the mouse, indicates quite strongly that activity of the embryonic genome certainly occurs during the preimplantation period and, moreover, that this activity is required for early development (see also *Engel* and *Franke*, this volume).

Acknowledgments: Work from the author's laboratory was supported by grants from the National Institutes of Health (HD 03132 and GM 19527). This paper was prepared while the author was on leave at the Imperial Cancer Research Fund Laboratories, London, as a Fellow of the John Simon Guggenheim Memorial Foundation.

References

Artzt, K., Dubois, P., Bennett, D., Condamine, H., Babinet, C., Jacob, F.: Surface antigens common to mouse cleavage embryos and primitive teratocarcinoma cells in culture. Proc. nat. Acad. Sci. (Wash.) 70, 2988-2992 (1973).

Brinster, R.L.: Parental glucose phosphate isomerase activity in three-day mouse embryos. Biochem. Genet. 9, 187-191 (1973).

Chapman, V.M., Wudl, L.: The expression of β-glucuronidase during embryogenesis, Third International Isozyme Conference. In: Isozymes III. Developmental Biology (ed. *C.L. Markert*), pp. 57-65. New York-San Francisco-London: Academic Press 1974.

Ellem, K.A.O., Gwatkin, R.B.L.: Patterns of nucleic acid synthesis in the early mouse embryo. Develop. Biol. 18, 311-330 (1968).

Engel, W., Franke, W.: Maternal storage in the mammalian oocyte. Curr. Topics in Pathology (eds. *E. Grundman, W.H. Kirsten*). 62, 29-52 (1976).

Epstein, C.J.: Phosphoribosyltransferase activity during early mammalian development. J. Biol. Chem. 245, 3289-3294 (1970).

Epstein, C.J.: Expression of the mammalian X-chromosome before and after fertilization. Science 175, 1467-1468 (1972).

Epstein, C.J.: Gene expression and macromolecular synthesis during preimplantation embryonic development. Biol. Reprod. 12, 82-105 (1975).

Epstein, C.J., Daentl, D.L.: Precursor pools and RNA synthesis in preimplantation mouse embryos. Dev. Biol. 26, 517-524 (1971).

Epstein, C.J., Daentl, D.L.: Molecular events in preimplantation mammalian development. In: Cell Differentiation (eds. *R. Harris, P. Allin, D. Viza*), pp. 69-72. Copenhagen: Munksgaard 1972.

Epstein, C.J., Smith, S.A.: Aminoacid uptake and protein synthesis in preimplantation mouse embryos. Dev. Biol. 33, 171-184 (1973).

Epstein, C.J., Smith, S.A.: Electrophoretic analysis of proteins synthesized by preimplantation mouse embryos. Dev. Biol. 40, 233-244 (1974).

Golbus, M.S., Calarco, P.G., Epstein, C.J.: The effects of inhibitors of RNA synthesis (α-amanitin and actinomycin D) on preimplantation mouse embryogenesis. J. exp. Zool. 186, 207-216 (1973).

Golbus, M.S., Epstein, C.J.: Effect of 5-bromodeoxyuridine on preimplantation mouse embryo development. Differentiation 2, 143-149 (1974).

Gross, P.R.: The control of protein synthesis in embryonic development and differentiation. Curr. Top. dev. Biol. 2, 1-46 (1967).

Kramen, M.A., Biggers, J.D.: Uptake of tricarboxylic acid cycle intermediates by preimplantation mouse embryos *in vitro*. Proc. nat. Acad. Sci. (Wash.) 68, 2656-2659 (1971)

Manes, C.: The participation of the embryonic genome during early cleavage in the rabbit. Dev. Biol. 32, 453-459 (1973).

Schultz, G., Manes, C., Hahn, W.E.: Synthesis of RNA containing polyadenylic acid sequences in preimplantation rabbit embryos. Dev. Biol. 30, 418-426 (1973a).

Schultz, G.A., Manes, C., Hahn, W.E.: Estimation of the diversity of transcription in early rabbit embryos. Biochem. Genet. 9, 247-259 (1973b).

Woodland, H.R., Graham, C.F.: RNA synthesis during early development of the mouse. Nature (Lond.) 221, 327-332 (1969).

Formation of the Blastocyst: Determination of Trophoblast and Embryonic Knot *

H.-W. DENKER

I. Introduction

In mammalian development, the earliest apparent differentiation of cells into two distinct lines manifests itself in the early blastocyst, when trophoblast and embryonic knot ("inner cell mass") become distinguishable. Recently more and more attention is being paid to this process by developmental biologists, geneticists, cell biologists, reproductive biologists, and oncologists. This is apparently for one of the following reasons: (1) Blastocyst formation seems to result from the first process of determination in mammalian ontogeny, which means that the genome has not been under the influence of any determinative stimuli before. (2) The system appears to be far less complex than later stages so that it may be especially suitable for experimentation. (3) Determination of trophoblast, a tissue of some peculiar properties, may exhibit interesting specific features. (4) Trophoblast is in the focus of interest of reproductive biologists as well as oncologists because of its role in mediating contact between embryo and mother, and because of its invasive growth.

Attempts of investigators to elucidate the mechanism of determination of trophoblast and embryonic knot led them to design elegant experiments which, in turn, resulted in formula-

* Dedicated to Professor Dr. F. Seidel.

tion of different theories to be described in this paper. Excellent reviews of part of the arguments summarized here have been given by *Mulnard* (1966), *Seidel* (1969), *Graham* (1971), *Gardner* (1973), and *Herbert* and *Graham* (1974).

Terminology

"Embryonic knot": In the present paper, we will use this term rather than the more commonly used term "inner cell mass" because the latter would fit only one of the theories to be described here, and because it would not be suitable for discussion of a possible onset of determination before certain cells acquire an inside position.

"Polarity": The term is being used in a general sense not referring specifically to polar organization along the animal-vegetal axis.

II. Trophoblast and Embryonic Knot Representing Distinct Populations of Cells (Differentiation of Trophoblast and Embryonic Knot)

Considering the process of blastocyst formation to be connected with a process of cell differentiation would require data showing that trophoblast and embryonic knot (or at least one of both) are in fact composed of distinctly differentiated cells. The following observations give support to this view:

1. Trophoblast and embryonic knot cells, when isolated from 3 1/2-day mouse blastocysts, differ in their ability to induce a decidual reaction in the pseudopregnant uterus: trophoblast cells do induce this reaction, whereas embryonic knot cells do not (*Gardner*, 1971, 1972a).

2. The same isolated trophoblast and embryonic knot cells also differ in their tendency to stick together and to form a common structure in vitro: embryonic knot cells do so, whereas trophoblast cells stay apart; the latter form fluid-filled vesicles, whereas the former produce only solid cell clusters. Embryonic knot cells, when injected into the cavity of another blastocyst, will become integrated and form part of the body of the embryo, even rat embryonic knot cells injected into a mouse blastocyst: this is not the case with likewise injected trophoblast cells (*Gardner*, 1971, 1972a; *Gardner* and *Johnson*, 1973).

3. Trophoblast cells of blastocysts are connected with each other by junctional complexes. Well-developed junctions can already be seen between blastomeres in the outer layer of cleavage stages, i.e., between presumptive trophoblast cells, whereas between inner cells of morulae or between trophoblast and embryonic knot of blastocysts they are more rare and remain primitive. Differences in density of the cytoplasm and in number as well as structure of certain organelles have also been described (rat: *Schlafke* and *Enders*, 1967; *Dvořak*, 1971; mouse: *Calarco* and *Brown*, 1969; rabbit: *Hessedahl,* 1971).

4. Mitosis rate seems to be different in both types of cells, the embryonic knot showing the higher values. There is already a difference between inside and outside cells of morulae

as judged from combination of cell number and ^{3}H-thymidine incorporation studies (*Barlow* et al., 1972).

5. In histochemical studies, differences between trophoblast and embryonic knot have been found relatively often, e.g. differences in phosphatase activity. Problems of interpretation of histochemical findings are in part discussed elsewhere (cf. also *Denker*, 1970).

6. Differences in cell surface properties of trophoblast and embryonic knot are possibly indicated by the fact that, in mouse blastocysts, both types of cells exhibit slightly different susceptibility to lysis by cytotoxic antisera (*Moskalewski* and *Koprowski*, 1972); furthermore, certain viruses injected into the blastocyst develop in the trophoblast but not in the embryonic knot (*Glass* et al., 1974).

7. Evidence has been presented for the expression of different esterase isoenzymes in trophoblast and embryonic knot of mouse blastocysts. The trophoblast-type isoenzyme A was detectable already in the late morula; this might be the earliest well-established biochemical criterium for beginning differentiation of trophoblast (*Sherman*, 1972).

The above mentioned data give evidence for morphologic, biochemical, and physiologic differences between trophoblast and embryonic knot. This may indicate that a process of differentiation has taken place, if this term is used in a merely descriptive sense – differentiation meaning the establishment of differential properties of cells. Direct evidence for differential gene activity of trophoblast and embryonic knot cells, however, is still lacking but might be found in the near future because it is already established that the genome is active in preimplantation embryos: Some indirect evidence is derived from investigations on synthesis of different classes of RNA and on changes in enzyme activities during preimplantation (although technical problems of determination of intracellular pools of precursors are not completely solved, and changes in enzyme activity do not necessarily reflect changes in genetic activity) (for review see *Church* and *Schultz*, 1974; *Graham*, 1973; *Woodland* and *Graham*, 1969). Effects of actinomycin D and α-amanitin demonstrate that RNA synthesis is indeed required for cleavage and blastocyst formation (*Golbus* et al., 1973; *Manes*, 1973). Direct evidence for genetic activity is derived from the fact that t^{12}/t^{12} homozygous mouse embryos die at the late morula stage (*Mintz*, 1964a) (apparently after onset of trophoblast differentiation, *Hillman* et al., 1970), and by the finding that the paternal phenotype of glucose phosphate isomerase isoenzymes is expressed in the blastocyst stage (*Chapman* et al., 1971), and the information for hypoxanthine-guanine phosphoribosyltransferase is apparently transcribed as early as the morula stage (*Epstein*, 1972).

Of the two classes of cells, trophoblast and embryonic knot, trophoblast gives better evidence that it undergoes real differentiation until the blastocyst stage: The trophoblast develops junctional complexes, secretes blastocyst fluid, is able to induce a decidual reaction, produces a specific esterase isoenzyme (see above) and a protease or protease activator (*Denker*, 1971a, 1974). The embryonic knot cells, on the contrary, seem to remain in a more primitive state (cf. also *Gardner*, 1971).

III. Determination of Trophoblast and Embryonic Knot

1. Theory A: Determination Depending on the Position of Blastomeres ("Inside-Outside Model") (Fig. 1A)

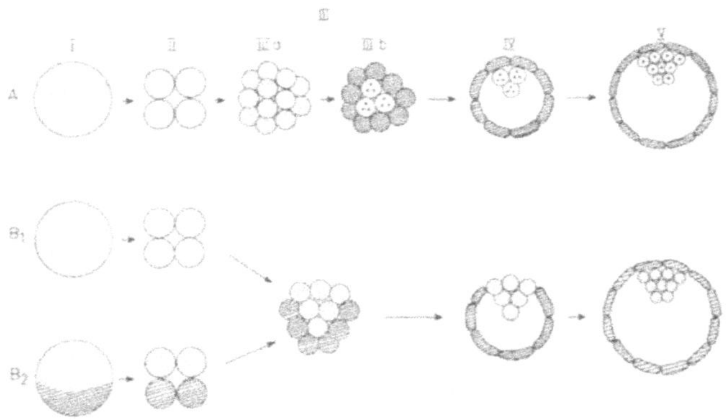

Fig. 1. Diagram to illustrate the two theories on determination of trophoblast and embryonic knot. Theory A (inside-outside model): In the beginning, the developmental potentials of all blastomeres are equal (A II, IIIa). When, in the course of cleavage, some cells become completely surrounded by others, these inside cells are being determined to form embryonic knot (inner cell mass) (A IIIb); the other (outside) blastomeres will differentiate into trophoblast.
Theory B: Determination depends on localized factors of polar (bilateral) distribution, which, in this diagram, are assumed to be trophoblast-determining. These factors are either (B_2) localized in a certain area of egg cytoplasm (B_2 I) and become segregated during cleavage so that they will be found only in certain blastomeres (B_2 II); or (B_1) factors are of unknown, maybe even exogenic, origin, but their action is nevertheless locally restricted. In both cases (B_1 and B_2), morulae show polarity (III) as a result of polar action of determining factors

According to this theory, there are no differences between blastomeres either in their status of determination or in their developmental potentials, until one or several of them become completely surrounded by others, i.e., in or after the eight-cell stage. The completely surrounded cells will form the embryonic knot (inner cell mass), whereas the whole of the outer layer will become trophoblast. The ability to differentiate into trophoblast is inherent to all blastomeres at the beginning; the information (the determinative stimulus) to form, instead, embryonic knot is provided by the specific milieu "inside" a morula or blastocyst (*Tarkowski* and *Wróblewska*, 1967).

Observations supporting theory A have mostly been made in the course of experimentation in the mouse. Two or more cleavage stage mouse embryos up to the morula stage can be fused together to form one single chimeric blastocyst developing into one single embryo (*Tarkowski*, 1961, 1965; *Mintz*, 1962, 1965). This demonstrates a remarkable regulative capacity of the early mammalian embryo. Even rat and mouse morulae can be combined this way to form interspecific chimeras (*Stern*, 1973; *Zeilmaker*, 1973). If, in these embryos, future trophoblast and embryonic knot cells (or one of both groups) had already been determined in a fixed and unchangeable manner, only a specific sorting out of cells during fusion could explain the formation of one single chimeric blastocyst. At first, this type of explanation has in fact been given (*Tarkowski*, 1961). Labeling experiments, however, did not give evidence for any specific sorting-out process (Fig. 2) (*Mintz*, 1964b, 1965; *Hillman* et al., 1972).

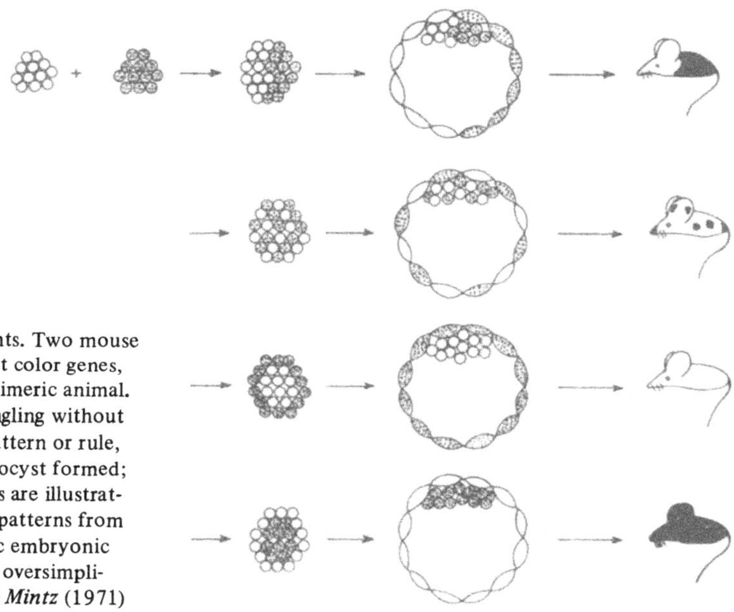

Fig. 2. Fusion experiments. Two mouse morulae, differing in coat color genes, are fused to form one chimeric animal. Blastomeres are intermingling without exhibiting any regular pattern or rule, as to be seen in the blastocyst formed; four extreme possibilities are illustrated. Resulting coat color patterns from blastocysts with chimeric embryonic knot (first two lines) are oversimplified; for exact details see *Mintz* (1971)

In fact, blastomeres put outside a morula tended to become incorporated into the trophoblast rather than the embryonic knot. Several variants of this experiment have been performed, all giving principally the same result. This obviously indicates that the developmental fate of blastomeres does, or at least can, depend on their position – inside or outside (Fig. 3) (*Hillman* et al., 1972). Note that in this type of experiment it is not known whether the blastomeres used for recombination are presumptive trophoblast or embryonic knot cells, so that conclusions can be drawn only from statistical analysis of proportions of labeled cells in the trophoblast (or embryonic knot) of the chimera.

Very impressive are experiments in which all blastomeres of a ^{3}H-thymidine-labeled 8-16 cell embryo were brought into an inside position by the way that this embryo was being surrounded by another 14 unlabeled embryos (Fig. 3) (*Hillman* et al., 1972). By this means, all cells of the labeled inside embryo could be forced, in some cases, to form part of the embryonic knot, even those which normally would have formed trophoblast.

Another way of labeling cells was used in experiments by *Wilson* et al. (1972) and *Stern* and *Wilson* (1972): silicone oil microdroplets were put into the cytoplasm. Peripheral cells of late morulae or early blastocysts of the mouse were labeled this way and the embryos fused with either identical stages or eight-cell embryos. In the resulting chimeras, part of the labeled outside cells were found in the embryonic knot. This again demonstrates that either the fate of outside (presumptive trophoblast) cells of the late morula/early blastocyst is not fixed yet, or that not all of these cells are presumptive trophoblast (cf. p. 72).

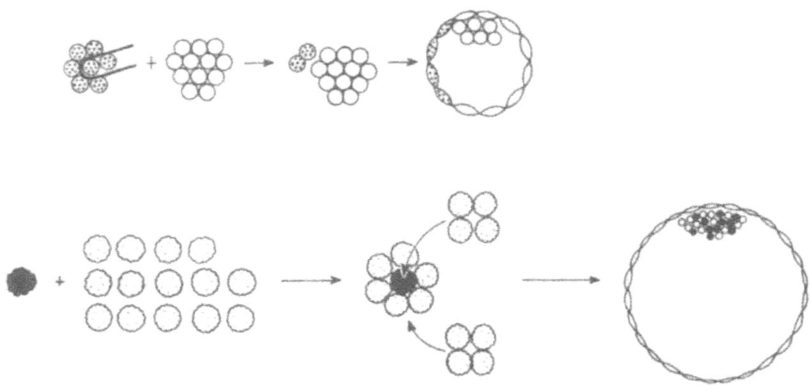

Fig. 3. Fusion experiments: Demonstration of decisive role of inside or outside position of blastomeres (after *Hillman* et al., 1972). Mouse embryos. Labeling was done with ^{3}H-thymidine.

Above: Two labeled blastomeres of 8-cell stage on outside of 8- to 16-cell unlabeled embryo. Position of labeled cells was determined in resulting blastocyst. In 40 experiments performed, 92% of all labeled cells were found in trophoblast.

Below: Labeled 8- to 16-cell embryo is being completely surrounded by 14 unlabeled embryos of same stage (note that each symbol here represents one whole embryo). Result is a giant blastocyst able to give rise to a well-shaped embryo. In 4 out of 7 experiments, labeled cells were found exclusively in embryonic knot

The marked ability of disaggregated mouse embryos (even blastocysts) to reaggregate and form regular blastocysts was also taken as an argument for developmental lability of blastomeres (*Stern*, 1972).

The type of experiments described above have been reviewed in detail by *Herbert* and *Graham* (1974).

Another line of evidence that theory A might be correct was derived from observations of coincidence; there was a marked increase of uridine incorporated into RNA in the stage when blastomeres begin to acquire an inside position. But note that in the mouse, all major classes of RNA are synthesized from the four-cell stage on, and on the other hand, the problem of interference of intracellular pools of precursors has not yet been solved (reviewed by *Graham*, 1973; *Church* and *Schultz*, 1974).

2. Theory B: Determination Depending on Polar Bilateral Organization of the Egg (Morula) (Fig. 1B_1, B_2)

According to this theory, determination does not depend on the inside or outside position of blastomeres but on morphogenetic factors restricted to a localized area in the egg or morula. In the most elaborate version of this theory (Fig. 1B_2), these factors are presumed to be present in a certain area of cytoplasm of the uncleaved egg; during cleavage, segregation occurs, and those blastomeres which receive part of this material will be determined to differentiate into a certain direction: e.g. if the factors are embryonic knot-determining, the progeny of these blastomeres will develop into embryonic knot (part of them may, in addition, also form trophoblast).

It is also possible to formulate a more general version of this theory which does not include polar organization of the uncleaved egg but only postulates that, during cleavage, the embryo acquires polarity due to unknown processes (governed e.g., by locally acting exogenic factors (Fig. $1B_1$).

In each case, according to this theory, the primary arrangement of presumptive trophoblast and embryonic knot cells in a cleavage stage embryo would exhibit polarity, depending on the eccentric localization of the determining factors. This is in clear contrast to the radial symmetry suggested by the inside-outside model (theory A, Fig. 1A).

Observations supporting theory B have been made in the course of experimentation and of morphologic and histochemical investigations in several species.

a) Experiments

One of the blastomeres of the rabbit two-cell embryo or three of the blastomeres of the four-cell stage were destroyed by pricking them with a needle, and the developmental potential of the surviving blastomere was followed (*Seidel*, 1952, 1956, 1960). The result, relevant to the problem discussed here, was that not all of these surviving blastomeres formed regular blastocysts: in a certain proportion of them (about 1:2 in case of experimentation in the two-cell stage) the embryonic knot was lacking and only trophoblast (and sometimes also entoderm) developed.

This experiment demonstrates limits of regulative capacities of the rabbit embryo. Referring also to comparative aspects based on numerous data from lower animals, *Seidel* formulated the theory that, in the mammalian egg cytoplasm, a specialized area exists (*Plasmatischer Faktorenbereich*, plasmatic field of factors), which is organized like a formative center (*Bildungszentrum*). Whereas all blastomeres primarily have the potential to form trophoblast, only those of them which receive, during cleavage divisions, part of the cytoplasmic field of factors will be able to differentiate, in addition, an embryonic knot (which means they have all the information available to form a whole blastocyst). The result of the deletion experiment (blastocyst or trophoblastic vesicle) would depend on whether the surviving blastomere by chance was one of those which do possess the factors, or one of those which do not (Fig. 4).

Comparable experiments in the mouse seemed to show the same trend, and at first the same interpretation has been given (*Tarkowski*, 1959, 1961). Later on, the results of studies of the developmental capacities of all blastomeres of four to eight cell embryos were given an interpretation in favor of the inside-outside model (theory A) (*Tarkowski* and *Wróblewska*, 1967) (see general discussion).

Labeling certain parts of cytoplasm by injecting silicone oil droplets revealed in the mouse egg, that there is no important spatial disturbance of the cytoplasmic pattern of the egg during cleavage: the cortical region of the egg being converted to the outer cells of the morula. (Nevertheless, fusion experiments done with these labeled embryos show that these cortical regions also can be forced to become part of the embryonic knot) (*Wilson* et al., 1972; *Stern* and *Wilson*, 1972).

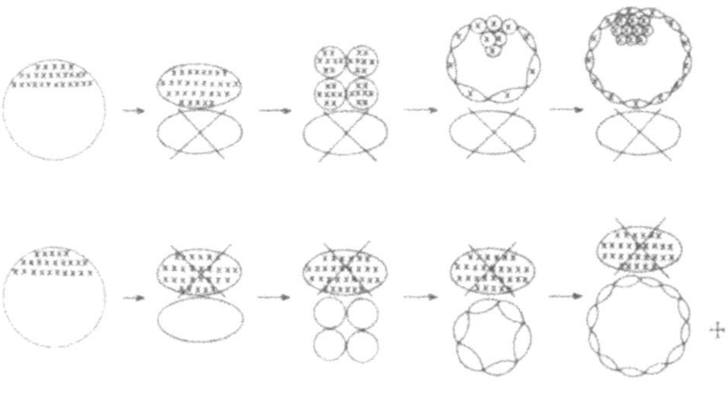

Fig. 4. Seidel's experiment in rabbit. One blastomere of the 2-cell stage is being destroyed (marked by large cross). The surviving blastomere will develop (after transfer of whole embryo into foster mother) into either a regular blastocyst with embryonic knot (*above*), or a trophoblastic vesicle without embryonic knot (*below*). Result is thought to reflect that the surviving blastomere either contained or lacked material of a cytoplasmic field of factors (*Bildungszentrum*) present in a restricted area of egg cytoplasm which provides information for determination of embryonic knot. Formation of trophoblast, on the contrary, is assumed to be a general ability common to all blastomeres (with or without *Bildungszentrum*)

The assumption that the cytoplasmic field of factors determines embryonic knot rather than trophoblast is not necessarily an integral part of theory A. The available experimental results do not completely rule out other interpretations like: Both embryonic knot – determining as well as trophoblast – determining fields of factors might exist at opposite poles of the egg. Or there might be only factors providing the information for determination of trophoblast; these factors might be organized like a field, and the embryonic knot be formed at the opposite pole. Available information does not yet allow one to decide which of these possibilities is the correct one. Because there is more evidence for the trophoblast than for the embryonic knot to undergo real differentiation during these early stages (see p. 61), we like to reillustrate *Seidel*'s experiment assuming that the cytoplasmic factors are trophoblast-determining (Fig. 5). Embryos lacking trophoblast (i.e., pure embryonic knots) (Fig. 5 I) were not found by *Seidel* (1960, 1969), or rarely found by *Tarkowski* and *Wróblewska* (1967). Apart from other possibilities, this could still be explained assuming that the field of factors is trophoblast-determining but extends over a rather large area so that even blastomeres of the opposite pole would still be able to form trophoblast (in addition to embryonic knot) (Fig. 6).

b) Morphologic and Histochemical Data

In a large series of papers *Dalcq* and *Mulnard* (*Dalcq*, 1951, 1954, 1955, 1962a, b, c, 1966; *Mulnard*, 1955, 1965; *Mulnard* and *Dalcq*, 1955) presented a number of data on histochem-

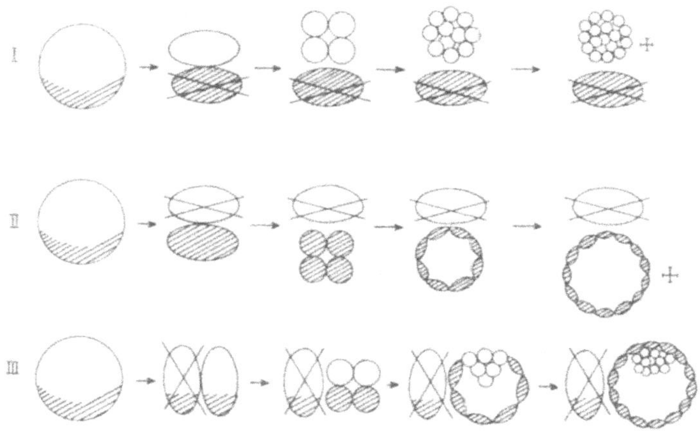

Fig. 5. Seidel's experiment in rabbit (cf. Fig. 4) redrawn assuming that the cytoplasmic field of factors is trophoblast-determining. First cleavage furrow can lie in different planes and either restrict this field to one of two blastomeres (I, II) or devide it (III). For descriptive purposes, the situation is oversimplified in this diagram illustrating mosaic-type reactions: Half-embryos which consist only of "trophoblast-factor cytoplasm" form only trophoblast (II), half-embryos without it form only embryonic knot (I), half-embryos with both types of cytoplasm form both types of cells, i.e. a whole blastocyst (III)

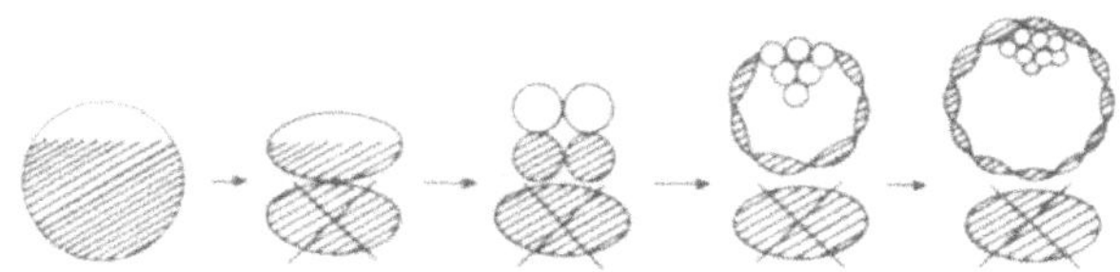

Fig. 6. Same as in Fig. 5 I, but assuming that the area which the trophoblast-determining cytoplasmic field of factors takes in the egg is larger. This could explain why pure embryonic knots (cf. Fig. 5 I) are never (or rarely) seen in the experiments. Another possible explanation for the same phenomenon is provided by theory described in Fig. 4 (legend)

ical differences between blastomeres, and correlated them with differences between trophoblast and embryonic knot seen in blastocysts. Most of this work was done in the rat, including a number of investigations in the mouse. The uncleaved egg was described to exhibit a plane of bilateral symmetry (most obvious in the rat) forming a certain angle with the animal-vegetal axis, the cytoplasm of the so-called dorsal side being especially rich in RNA (basophilic region, cf. *Jones-Seaton*, 1950; see also *De Geeter*, 1954) and exhibiting a characteristic diffuse type of acid phosphatase reaction (*Mulnard*, 1955, 1965). In cleavage stages, these histochemical characteristics were found to be restricted to certain blastomeres that finally will form the embryonic knot after becoming enveloped by the presumptive trophoblast cells (cf. e.g. *Mulnard*, 1966, Fig. 2). As a submicroscopic equivalent of the basophilic region, *Krauskopf* (1968) described in the rabbit egg an area rich in polyribosomes and poor in other organelles. There is no report on comparable observations by other electron microscopists.

The conclusions drawn from the above-mentioned investigations are not widely accepted today (cf. *Tarkowski* and *Wróblewska*, 1967; *Herbert* and *Graham*, 1974). Reinvestigations of the phosphatase distribution using azo-dye methods instead of Gomori-type reactions failed to confirm differences of enzyme reaction between blastomeres in the rat (*Rodé* et al., 1968) or mouse (*Denker*, unpublished), although in the hamster they did show the more intense reaction of the embryonic knot (*Ishida*, 1972).

If there are histochemical differences between blastomeres, it is most relevant for the discussion of theories A and B whether the arrangement is similar to Figure 1 A IIIb or Figure 1 B III. Diagrams showing the latter type of blastomere arrangement can often be found in *Dalcq*'s and *Mulnard*'s publications (e.g. *Mulnard*, 1966, Fig. 2), but unfortunately, photographs documenting this unequivocally are rare because of technical difficulties with photography of whole mounts.

Table 1. List of references showing polar arrangement of two groups of blastomeres in eutherian mammals.

Species	References
Rabbit	*van Beneden* (1880) Pl. IV Fig. 1-3, 7
Rabbit	*Assheton* (1895) Pl. 14 Fig. 18, 20; Pl.15 Fig. 22
Bat	*Duval* (1895) Pl. III Fig. 9, 14, 15, 17, 18, 20, 21; Fig. IX p. 140; Fig. X p. 151
Bat	*van Beneden* (1899) Fig. 1 p. 310
Bat	*van Beneden* (1911) Pl. II Fig. 25, 29; Pl. III Fig. 37, 40
Bat	*van Beneden* and *Julin* (1880) Pl. XXIII Fig. 5, 6
Bat	*Wimsatt* (1944) Fig. 23 p. 404/405
Mole	*Heape*(1886) Pl. XI Fig. 20
Sheep	*Assheton* (1898/99) Pl. 15 Fig. 7, 8; Pl. 18; Text-Fig. p. 222
Goat	*Amoroso* et al. (1942) Text-Fig. 4 p. 388; Text-Fig. 5 p. 390
Pig	*Heuser* and *Streeter* (1929) Fig. 5 p. 16; Fig. 6 p. 17; Pl. 12
Rat, mouse	*Skalko* (1971) Fig. 15-1 d p. 242
	Dalcq; Mulnard (various papers, see list of references)

In the older morphologic literature, illustrations (again mostly drawings) that show such a polar arrangement of blastomeres can be found more often (Table 1). They are mostly based on purely morphologic criteria, like cell size. This often seems to include a lot of interpretation, especially when blastomeres of intermediate size exist but are nevertheless put in one or the other of the two categories of cells (*Heuser* and *Streeter*, 1929; cf. their Fig. 4, p. 14, with Figs. 5 and 6, pp. 16 and 17).

Recently, *Denker* (1970, 1971b, 1972) described a similar polar arrangement of two distinct types of blastomeres in rabbit embryos. After formol-alcohol-acetic acid fixation, which was thought to either chemically modify or extract certain fractions of proteins

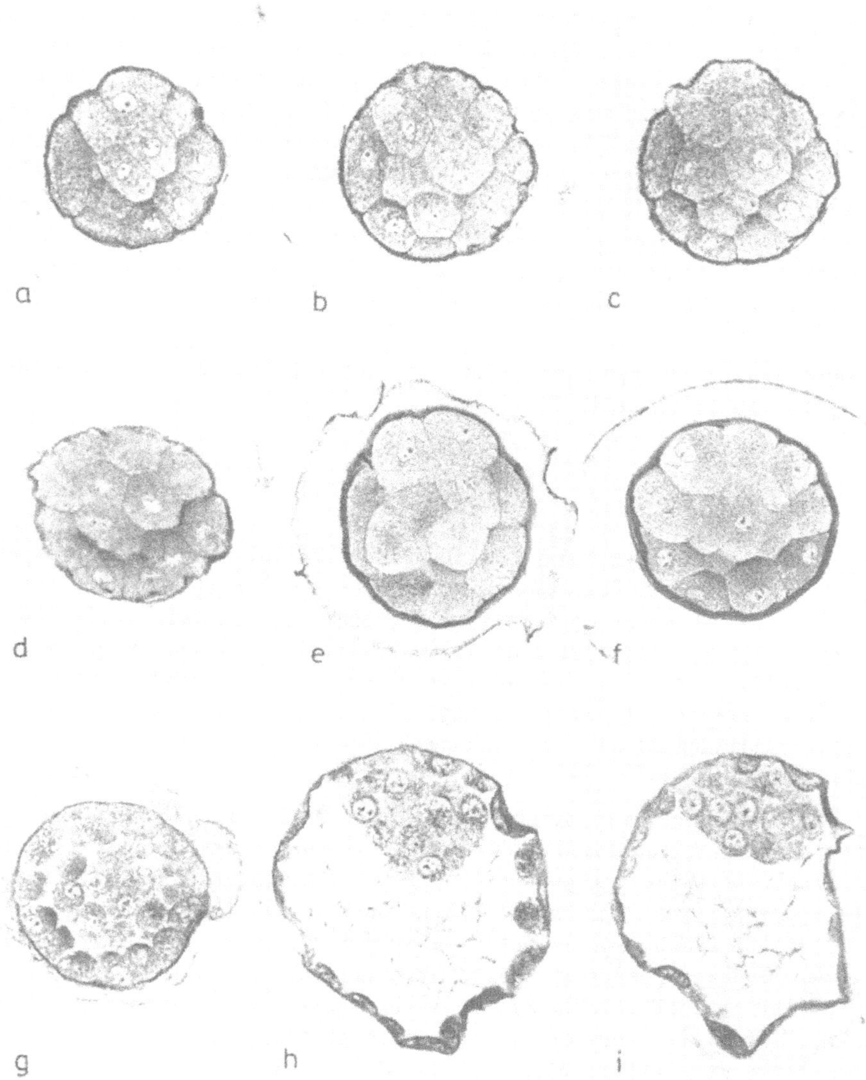

Fig. 7. Aspects of polar arrangement of two groups of blastomeres in the rabbit. Embryos were fixed with formol-alcohol-acetic acid, paraffin sections were taken, proteins were stained with Hg-bromphenol blue. x 260. a-c: 54 h p.c., three sections from same embryo; d: 54 h p.c.; e, f: 63 h p.c.; g: 76 h p.c., blastocyst cavity just appearing (cleft); h, i: 80 h p.c., early blastocyst.

Two categories of blastomeres can be distinguished by intensity of cytoplasmic staining: 1. Lighter stained, not polarized cells; in early blastocyst, light cells form embryonic knot. 2. Darker stained cells which often show very obvious maximum of stain uptake in parts of cytoplasm directed toward center of egg. When cavitation begins, they can be identified as prospective trophoblast cells (g). Trophoblast of early blastocysts also stains more intensely than embryonic knot (h, i). Note that in cleavage stages trophoblast-type cells form a single-layered cap which surrounds other blastomeres only incompletely (polarity!), but area taken in different eggs is different (a-f)

or nucleoproteins (*Denker*, 1972, Aussprache), there were very obvious differences in histochemically detectable concentrations of protein and RNA in the cytoplasm of different groups of blastomeres. The arrangement of the two categories of blastomeres corresponded to the illustration given in Figure 1 B III, and remained constant over a long series of cleavage stages, from the eight-cell stage to the late morula, i.e., in this species over a period of at least three cell cycles (Fig. 7). Although precautions are necessary when conclusions are to be drawn from histochemical observations (for discussion cf., e.g., *Denker*, 1970), it seemed to be possible to follow the fate of the protein-rich (darker) cells to become trophoblast, and the lighter stained cells to give rise to the embryonic knot. The presumptive trophoblast cells seemed to envelope (in an epiboly-like way) the presumptive embryonic knot cells, which acquired a completely inside position only at or even after the time when the blastocyst cavity appeared.

In the mouse and hamster, however, the same technique failed to show any clear differences between blastomeres (*Denker*, 1972).

IV. General Discussion

The two controversial theories A and B described above are both based mainly on experimental data, but the types of experiments are different: transplantation experiments in the case of the inside-outside model (theory A) (fusion of morulae; transfer of labeled single blastomeres), isolation experiments (destruction of all blastomeres except one; disaggregation of blastomeres) in the case of theory B (including also histochemical findings). We shall discuss now which type of experiment might give the more relevant answer to the question concerning the stages and basic mechanisms of determining trophoblast and embryonic knot.

Seidel (1953, p. 91) gives the following *definition of determination: 'Für eine Aufgabe determiniert' kann lediglich ausdrücken: 'Der Keimteil besitzt in seinem derzeitigen Zustand unter den vorhandenen Bedingungen eine bestimmte Entwicklungsbefähigung und kann sie selbständig verwirklichen.' Offen muß bleiben, welche nicht genannten Fähigkeiten noch außerdem im Keim schlummern und durch Änderung der Bedingungen erweckt werden können oder spontan in ihm hervortreten.* (" 'To be determined' cannot mean anything more than: 'This specific part of the embryo has, in its present state and under the given conditions, a defined developmental potential, and is able to express it in an independent way.' This does not say anything about additional potentials which the embryo may have and which could manifest themselves either when conditions are changed, or even spontaneously." Translation by author).

Spemann (1936, p. 23) similarly uses the ability of cells to differentiate independently as the most important criterium: *Wenn ein Keimteil die Ursachen einer bestimmt gerichteten Weiterentwicklung in sich selbst trägt, so kann man sagen, daß er zu seinem Schicksal bestimmt, 'determiniert', ist. Jedenfalls kann man mit Lillie (1929) den Begriff der Determination so fassen, daß man die Selbstdifferenzierungsfähigkeit zu seinem Kriterium macht.* ("Whenever a certain part of an embryo carries in itself the causes for its development into a certain direction, it is possible to say that this part is 'determined' for its fate. At least it is possible to define the term determination that way, using the ability for self-differentiation as a criterium, according to *Lillie* (1929)." Translation by author).

It is not necessary to discuss here the concept of "self-determination" which itself can be criticized for being, in its strict form, too narrow for most systems. The important point is that the ability to differentiate after isolation, which can be proved experimentally, is being used as a criterium. If this definition of determination is accepted, it follows that isolation experiments should give better information about the determined or nondetermined status of a cell than transplantation experiments can do. In the latter type of experiments, the cells are brought under the influence of different parts of the embryo (or other tissues), and the original state of the transplant and the conditions are drastically changed. *Spemann*, who has done a lot of transplantation experiments, comments (1936, p. 31): *"So wird Transplantation im neuen Gewebsverband nur dann sichere Auskunft geben können, wenn Selbstdifferenzierung stattfindet, wenn also die Determination des Implantats genügend befestigt ist, um sich auch gegen einen etwaigen Einfluß der Umgebung durchzusetzen. Der erste Eintritt der Determination wird sich dagegen nur bei völliger Isolierung erkennen lassen."* (Transplantation . . . "will give a clear answer only in case of self-differentiation, i.e., when determination of the implant is stabile enough to dominate over possible influences of the surrounding tissues. The very beginning of determination can only be recognized in case of complete isolation." Translation by author).

According to this, the results of the described isolation experiments which favor theory B would seem more relevant for the discussion of mechanisms involved in determination of trophoblast and embryonic knot. The transplantation experiments described on pp. 62 ff., on the other hand, demonstrate the high regulative capacity of these early embryonic states. When, for example, all blastomeres of a morula, even those which normally would have developed into trophoblast, can be forced to form part of the embryonic knot (p. 63, Fig. 3), this shows in an impressive way the flexibility of the system and that it does not exhibit features of a mosaic. But is seems questionable if this type of experiment can uncover whether cells are already *inclined* (but not irreversibly switched yet) to form trophoblast, which isolation experiments possibly do reveal.

Often a different definition of cell determination is being used, e.g. by *Herbert* and *Graham* (1974): "Cell determination is the process by which the developmental potential of a cell becomes limited during embryogenesis." Transplantation experiments can certainly uncover limited developmental potentials. This is definitely the case when the limitation became irreversible. But, we feel that this is a secondary process following the establishment of an inclination of cells to develop into one or the other direction.

Isolation experiments are also certainly problematic because they do change the state and conditions of cells, both due to the isolation procedure (e.g. pronase and EDTA-treatment, *Tarkowski* and *Wróblewska*, 1967) and to in vitro culture conditions. If part of the embryo is being destroyed but not removed (*Seidel*, 1960), it might influence the results of the experiment: when, in the amphibian embryo, one blastomere of a two-cell stage was destroyed but not removed, the surviving blastomere produced only a half embryo (*W. Roux*); on the contrary, when both blastomeres were completely separated, they both regulated and each one formed a whole twin embryo (*Spemann*, 1936, p. 11 ff.). It is therefore interesting that in *Tarkowski*'s and *Wróblewska*'s experiments (1967), the completely isolated blastomeres did develop into different forms of vesicles (blastocysts, "false blastocysts," trophoblastic vesicles), and some of them even formed only uncavitated masses of cells. The authors felt, however, that these differences might be due to differences in culture conditions. They concluded that: 1. At least in some eggs, all blastomeres have the potential to form vesicular structures (this view also being part of *Seidel*'s theory, see p. 65). 2. Incidence of differentiation into real blastocysts decreases from one-quarter blastomeres to one-eighth forms. *Tarkowski* and *Wróblewska* feel that most probably this is due to the fact that with advancing stage of development of the blastomeres at the time of their separation, the number of cells attained by them at the time of cavitation decreases. Consequently, the probability that cells become enveloped decreases. The authors imagine that up to the eight-cell stage, all blastomeres possess the ability to differentiate into trophoblastic direction, and development into embryonic knot cells is being triggered by an inside position (theory A).

Data presented by *Moore* et al. (1968) (rabbit) unfortunately do not contribute to the discussion of these two theories because an account of purely trophoblastic structures is not included.

As a different kind of approach, morphologic and histochemic analyses give additional information. By fixing the embryos it is intended to preserve certain characteristics of the momentary state of blastomeres without changing them by initiating regulatory processes as it probably happens in both transplantation and isolation experiments. The polar arrangement of different types of blastomeres as seen in the rabbit morula seems to form an argument for theory B (see p. 68 f.). This arrangement also provides important aspects for the interpretation of fusion experiments: if it is correct, as suggested by the histochemical findings, that only some but not all of the peripheral cells are determined to form trophoblast, then the fact that a proportion of labeled outside cells become included in the embryonic knot of the chimera (p. 63 f.) could reflect simply the yet undetermined state of some of the outside cells. This finding would then not contradict theory B anymore.

Theory B, in its elaborate form, implies that a specific area of the egg cytoplasm exerts a regulatory effect on gene activity. Influences of the egg cytoplasm on nuclei are in fact known from experiments in which egg nuclei were replaced by somatic cell nuclei (*Gurdon*, 1962; *Gurdon* and *Woodland*, 1969). There is some first evidence for the same phenomenon in the mouse egg (*Bernstein* and *Mukherjee*, 1972). Localization of factors of this type in specific areas of egg cytoplasm has been demonstrated in certain species (insect egg: *Seidel*, 1936).

A promising different approach is to study cell strains derived from preimplantation embryos, although there are certainly numerous differences between cultured cells and the original cells of the embryo. Trophoblast-resembling cells developed in vitro when isolated blastomeres of a stage as early as the four-cell stage were used (*Edwards*, 1964; *Cole* and *Paul*, 1965; *Cole* et al., 1965, 1966). It is hoped that additional information will come from investigations of teratomas derived from ectopically transplanted embryos or from unfertilized eggs in the ovary (*Evans*, 1972; *Damjanov* and *Solter*, 1974).

Finally, some comparative aspects shall be mentioned. In marsupials, cells are being determined to become embryonic disc (formative cells) without ever having been enclosed by other cells (Fig. 8) (*Hill*, 1911; *Hartman*, 1919). The same seems to hold true for at least one eutherian mammal: Hemicentetes (Tenrecoidea) (*Goetz*, 1937, 1939; *Bluntschli*, 1938) (Fig. 8). The inside-outside model (theory A) cannot be applied to these embryos because inside cells do not exist in these species before formation of the embryonic knot/disc. Probably the polarity of the blastocyst (embryonic disc, trophoblast) here results from a polarity of the uncleaved egg.

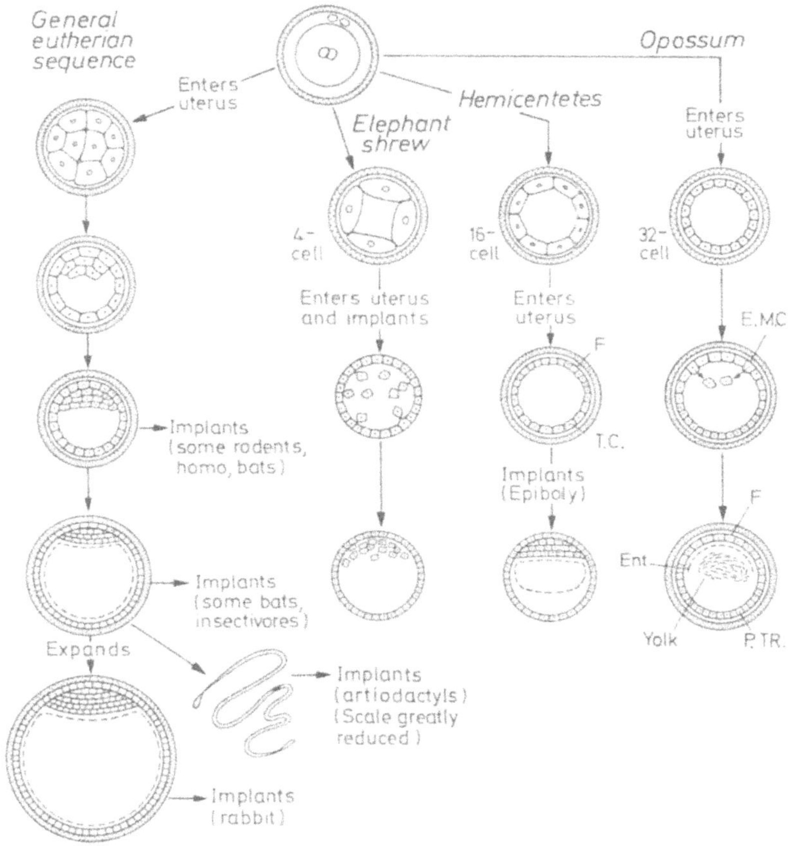

Fig. 8. Comparative aspects. In opossum and in Hemicentetes, cells being determined to form embryonic disc without having been in inside position. *E.M.C.:* entodermal mother cell; *Ent:* entoderm; *F:* formative cells; *P.TR.:* primitive trophoblast; *T.C.:* trophoblast cells. (For data on opossum and Hemicentetes, see *Hill*, 1911; *Hartman*, 1919; *Goetz*, 1937, 1939; *Bluntschli*, 1938) (from *Wimsatt*, 1975, by courtesy of author and of editor, Biology of Reproduction; slightly modified)

In conclusion, it appears well-established that the mammalian cleavage stage embryo possesses vast regulative capacities as impressively demonstrated by transplantation (fusion) experiments. The inside or outside position of blastomeres does influence their fate and can become decisive for their determination to form either trophoblast or embryonic knot. On the other hand, the egg does exhibit polarity, and blastomeres seem to be unequal, independent of their inside or outside position, as shown by their histochemical properties as well as their inclination to form only trophoblast or both trophoblast and embryonic knot. It is probable that this type of "preformation" is, in the beginning, weak and changeable, and can easily escape the experimentalist (cryptic preformation, *Graham*, 1971). The question remains open which type of experiment might be the most suitable to reveal physiologic in vivo mechanisms rather than in vitro regulations.

Acknowledgments. This review was written during the author's work at the Arbeitsgruppe Prof. Dr. *G.H.M. Gottschewski* am Max-Planck-Institut für Immunbiologie, Freiburg (West Germany).

The author wishes to thank Mrs. *M. Polack* for preparing the diagrams, and Dr. *W.A. Wimsatt* for the permission to use his diagram on comparative aspects (Fig. 8). Own experiments included in this review were supported by the Deutsche Forschungsgemeinschaft (De 181/1-3).

References

Amoroso, E.C., Griffiths, W.F.B., Hamilton, W.J.: The early development of the goat (Capra hirsus). J. Anat. (Lond.) 76, 377-406 (1942).

Assheton, R.: A re-investigation into the early stages of the development of the rabbit. Quart. J. micr. Sci. 37, 113-164 (1895).

Assheton, R.: The segmentation of the ovum of the sheep, with observations on the hypothesis of a hypoblastic origin for the trophoblast. Quart. J. micr. Sci. 41, 205-262 (1898/99).

Barlow, P., Owen, D.A.J., Graham, C.: DNA synthesis in the preimplantation mouse embryo. J. Embryol. exp. Morph. 27, 431-445 (1972).

Beneden, E. van: Recherches sur l'embryologie des mammifères. La formation des feuillets chez le Lapin. Arch. Biol. 1, 137-224 (1880).

Beneden, E. van: Recherches sur les premiers stades du développement du Murin (Vespertilio murinus). Anat. Anz. 16, 305-334 (1899).

Beneden, E. van: Recherches sur l'embryologie des mammifères. De la segmentation, de la formation de la cavité blastodermique et de l'embryon didermique chez le Murin. Arch. Biol. 26, 1-63 (1911).

Beneden, E. van, Julin, C.: Observations sur la maturation, la fécondation et la segmentation de l'oeuf chez les cheiroptères. Arch. Biol. 1, 551-571 (1880).

Bernstein, R.M., Mukherjee, B.B.: Control of nuclear RNA synthesis in 2-cell and 4-cell mouse embryos. Nature (Lond.) 238, 457-459 (1972).

Bluntschli, H.: Le développement primaire et l'implantation chez un Centetiné (Hemicentetes). C.R. Ass. Anat. 44, 39-46 (1938).

Calarco, P.G., Brown, E.H.: An ultrastructural and cytological study of preimplantation development of the mouse. J. exp. Zool. 171, 253-284 (1969).

Chapman, V.M., Whitten, W.K., Ruddle, F.H.: Expression of paternal glucose phosphate isomerase-1 (GPI-1) in preimplantation stages of mouse embryos. Develop. Biol. 26, 153-158 (1971).

Church, R.B., Schultz, G.A.: Differential gene activity in the pre- and postimplantation mammalian embryo. Curr. Top. Develop. Biol. 8, 179-202 (1974).

Cole, R.J., Edwards, R.G., Paul, J.: Cytodifferentiation in cell colonies and cell strains derived from cleaving ova and blastocysts of the rabbit. Exp. Cell Res. 37, 501-504 (1965).

Cole, R.J., Edwards, R.G., Paul, J.: Cytodifferentiation and embryogenesis in cell colonies and tissue cultures derived from ova and blastocysts of the rabbit. Develop. Biol. 13, 385-407 (1966).

Cole, R.J., Paul, J.: Properties of cultured preimplantation mouse and rabbit embryos, and cell strains derived from them. In: Preimplantation Stages of Pregnancy (eds. *G.E.W. Wolstenholme, M. O'Connor*), pp. 82-122. London: Churchill 1965.

Dalcq, A.M.: New descriptive and experimental data concerning the mammalian egg, principally of the rat (With a specification of a previous interpretation). Proc. kon. ned. Akad. Wet., Ser. C, 54, 351-363, 364-372, 469-479 (1951).

Dalcq, A.M.: Nouvelles données structurales et cytochimiques sur l'oeuf des mammifères. Rev. gén. Sci. Pures. Appl. 61, 19-41 (1954).

Dalcq, A.M.: Processes of synthesis during early development of rodent's eggs and embryos. In: Studies on Fertility (ed. *R.G. Harrison*) 7, 113-122 (1955).

Dalcq, A.M.: Evolution de l'organisation morphogénetique dans l'oocyte chez le Rat et la Souris. Verh. Anat. Ges., Anat. Anz. Suppl. 109, 373-382 (1962a).

Dalcq, A.M.: Localisations et evolution des phosphatases aux premiers stades du développement. Bull. Acad. roy. Med. Belg., VIIe Série, 2, 573-610 (1962b).

Dalcq, A.M.: Etudes cyto-enzymologiques sur les oeufs vivants de Souris incubés en présence d'ATP et d'autres mononucléotides. Arch. Biol. 73, 405-444 (1962c).

Dalcq, A.M.: Détection des enzymes de déphosphorylation dans les oeufs de rat et de souris fixés au formol et traités in toto. Arch. Biol. 77, 205-344 (1966).

De Geeter, L.: Etudes sur la structure de l'oeuf vierge et les premiers stades du développement chez le Cobaye et le Lapin. Arch. Biol. 65, 363-436 (1954).

Denker, H.-W.: Topochemie hochmolekularer Kohlenhydratsubstanzen in Frühentwicklung und Implantation des Kaninchens. II. Beiträge zu entwicklungsphysiologischen Fragestellungen. Zool. Jahrb. Abt. Allg. Zool. Physiol. 75, 246-308 (1970).

Denker, H.-W.: Enzym-Topochemie von Frühentwicklung und Implantation des Kaninchens. III. Proteasen. Histochemie 25, 344-360 (1971a).

Denker, H.-W.: In discussion to: *J.G. Mulnard*: Manipulation of cleaving mammalian embryo . . . Advanc. Biosci. 6, 255-277 (1971b).

Denker, H.-W.: Furchung beim Säugetier: Differenzierung von Trophoblast- und Embryonalknotenzellen. Verh. Anat. Ges. 66, Anat. Anz. Suppl. 130, 267-272 (1972).

Denker, H.-W.: Trophoblastic factors involved in lysis of the blastocyst coverings and in implantation in the rabbit: observations on inversely orientated blastocysts. J. Embryol. exp. Morph. 32, 739-748 (1974).

Duval, M.: Etudes sur l'embryologie des Chéiroptères. J. Anat. Physiol. 31, 93-160 (1895).

Dvořak, M.: Submicroscopic cytodifferentiation. Ergebn. Anat. Entwickl.-Gesch. 45/4 (1971).

Edwards, R.G.: Cleavage of one- and two-celled rabbit eggs in vitro after removal of the zona pellucida. J. Reprod. Fertil. 7, 413-415 (1964).

Epstein, C.J.: Expression of the mammalian X chromosome before and after fertilization. Science 175, 1467-1468 (1972).

Evans, M.J.: The isolation and properties of a clonal tissue culture strain of pluripotent mouse teratoma cells. J. Embryol. exp. Morph. 28, 163-176 (1972).

Gardner, R.L.: Manipulations on the blastocyst. Advanc. Biosci. 6, 279-301 (1971).

Gardner, R.L.: An investigation of inner cell mass and trophoblast tissues following their isolation from the mouse blastocyst. J. Embryol. exp. Morph. 28, 279-312 (1972a).

Gardner, R.L.: Manipulation of development. In: Reproduction in Mammals, Book 2: Embryonic and Fetal Development (eds. *C.R. Austin, R.V. Short*), pp. 110-133. Cambridge University Press 1972b.

Gardner, R.L.: Differentiation of trophoblast in the early mammalian embryo. Res. Reprod. 5 No. 5, 1-2 (1973).

Gardner, R.L., Johnson, M.H.: Investigation of early mammalian development using interspecific chimaeras between rat and mouse. Nature (New Biol.) 246, 86-89 (1973).

Glass, R.H., Calarco, P.G., Lin, T.P., Florence, J., Oh, J.O.: Development of the mouse blastocyst following injection with Newcastle disease virus. Biol. Reprod. 10, 502-511 (1974).

Goetz, R.H.: Studien zur Placentation der Centetiden. II. Die Implantation und Frühentwicklung von Hemicentetes semispinosus (Cuvier). Z. Anat. Entwickl.Gesch. 107, 274-318 (1937).

Goetz, R.H.: On the early development of the Tenrecoidea (Hemicentetes semispinosus). Bio-Morphosis 1, 67-79 (1939).

Golbus, M.S., Calarco, P.G., Epstein, C.J.: The effects of inhibitors of RNA synthesis (α-amanitin and actinomycin D) on preimplantation mouse embryogenesis. J. exp. Zool. 186, 207-216 (1973).

Graham, C.F.: The design of the mouse blastocyst. Symp. Soc. exp. Biol. 25, 371-378 (1971).

Graham, C.F.: Nucleic acid metabolism during early mammalian development. In: The Regulation of Mammalian Reproduction (eds. *S.J. Segal* et al.), pp. 286-301. Springfield, Ill.: Charles C.Thomas 1973.

Gurdon, J.B.: The developmental capacity of nuclei taken from intestinal epithelium cells of feeding tadpoles. J. Embryol. exp. Morph. 10, 622-640 (1962).

Gurdon, J.B., Woodland, H.R.: The influence of the cytoplasm on the nucleus during cell differentiation, with special reference to RNA synthesis during amphibian cleavage. Proc. roy. Soc. B 173, 99-111 (1969).

Hartman, C.G.: Studies on the development of the opossum Didelphys virginiana L. III. Description of new material on maturation, cleavage and entoderm formation. IV. The bilaminar blastocyst. J. Morph. 32, 1-142 (1919).

Heape, W.: The development of the mole (Talpa europea), the ovarian ovum, and segmentation of the ovum. Quart. J. micr. Sci. 26, 157-174 (1886).

Herbert, M.C., Graham, C.F.: Cell determination and biochemical differentiation of the early mammalian embryo. Curr. Top. Develop. Biol. 8, 151-178 (1974).

Hesseldahl, H.: Ulstrastructure of early cleavage stages and preimplantation in the rabbit. Z. Anat. Entwickl. Gesch. 135, 139-155 (1971).

Heuser, C.H., Streeter, G.L.: Early stages in the development of pig embryos, from the period of initial cleavage to the time of the appearance of limb-buds. Contr. Embryol. Carneg. Inst. 20, 1-29, (1929).

Hill, J.P.: The early development of the marsupialia, with special reference to the native cat (Dasyurus viverrinus). (Contributions to the embryology of the marsupialia, IV.). Quart. J. micr. Sci. 56, 1-134 (1911).

Hillman, N., Hillman, R., Wileman, G.: Ultrastructural studies of cleavage stage t^{12}/t^{12} mouse embryos. Amer. J. Anat. 128, 311-340 (1970).

Hillman, N., Sherman, M.I., Graham, C.: The effect of spacial arrangement on cell determination during mouse development. J. Embryol. exp. Morph. 28, 263-278 (1972).

Ishida, K.: Enzymohistochemical studies of differentiated 8-16 cell eggs of the hamster. Jap. J. Animal Reprod. 18, 105-109 (1972).

Jones-Seaton, A.: Etude de l'organisation cytoplasmique de l'oeuf des rongeurs, principalement quant à la basophilie ribonucléique. Arch. Biol. 61, 291-444 (1950).

Krauskopf, C.: Elektronenmikroskopische Untersuchungen über die Struktur der Oozyte und des 2-Zellenstadiums beim Kaninchen. II. Blastomeren. Z. Zellforsch. 92, 296-312 (1968).

Manes, C.: The participation of the embryonic genome during early cleavage in the rabbit. Develop. Biol. 32, 453-459 (1973).

Mintz, B.: Formation of genotypically mosaic mouse embryos. Amer. Zool. 2, 432 (1962).

Mintz, B.: Gene expression in the morula stage of mouse embryos, as observed during development of t^{12}/t^{12} lethal mutants in vitro. J. exp. Zool. 157, 267-271 (1964a).
Mintz, B.: Synthetic processes and early development in the mammalian egg. J. exp. Zool. 157, 85-100 (1964b).
Mintz, B.: Experimental genetic mosaicism in the mouse. In: Preimplantation Stages of Pregnancy (eds. *G.E.W. Wolstenholme, M. O'Connor*), pp. 194-216. London: Churchill 1965.
Mintz, B.: Clonal basis of mammalian differentiation. Symp. Soc. exp. Biol. 25, 345-370 (1971).
Moore, N.W., Adams, C.E., Rowson, L.E.A.: Developmental potential of single blastomeres of the rabbit egg. J. Reprod. Fertil. 17, 527-531 (1968).
Moskalewski, S., Koprowski, H.: Presence of egg antigen in immature oocytes and preimplantation embryos. Nature (Lond.) 237, 167-168 (1972).
Mulnard, J.: Contribution à la connaissance des enzymes dans l'ontogénèse. Les phosphomonoestérases acide et alkaline dans le développement du Rat et de la Souris. Arch. Biol. 66, 525-685 (1955).
Mulnard, J.: Aspects cytochimiques de la régulation in vitro de l'oeuf de souris après déstruction d'un des blastomeres du stade II. I. La phosphomonoestérase acide. Bull. Acad. roy. Med. Belg., Series 2, 5, 31-67 (1965).
Mulnard, J.: Les mécanismes de la regulation aux premiers stades du développement des mammifères. Bull. Soc. zool. Fr. 91, 253-277 (1966).
Mulnard, J.: Analyse microcinematographique du développment de l'oeuf de souris du stade II au blastocyste. Arch. Biol. 78, 107-138 (1967).
Mulnard, J., Dalcq, A.M.: Les polysaccharides dans le développement de l'oeuf tubaire du Rat. C.R. Soc. Biol. (Paris) 149, 836-839 (1955).
Rodé, B., Damjanov, I., Škreb, N.: Distribution of acid and alkaline phosphatase activity in early stages of rat embryos. Bull. Sci., Cons. Acad. RSF Yougoslavie, Sect. A, 13, 304 (1968).
Schlafke, S., Enders, A.C.: Cytological changes during cleavage and blastocyst formation in the rat. J. Anat. (Lond.) 102, 13-32 (1967).
Seidel, F.: Entwicklungsphysiologie des Insektenkeimes. Verh. dtsch. zool. Ges., Freiburg p. 291-336 (1936).
Seidel, F.: Die Entwicklungspotenzen einer isolierten Blastomere des Zweizellenstadiums im Säugetierei. Naturwissenschaften 39, 355-356 (1952).
Seidel, F.: Entwicklungsphysiologie der Tiere. I. Ei und Furchung. Sammlung Göschen Band 1162. Berlin: Walter de Gruyter & Co. 1953.
Seidel, F.: Nachweis eines Zentrums zur Bildung der Keimscheibe im Säugetierei. Naturwissenschaften 43, 306-307 (1956).
Seidel, F.: Die Entwicklungsfähigkeiten isolierter Furchungszellen aus dem Ei des Kaninchens Oryctolagus cuniculus. Wilhelm Roux' Arch. Entwicklungsmech. Organismen 152, 43-130 (1960).
Seidel, F.: Entwicklungspotenzen des frühen Säugetierkeimes. Arbeitsgemeinschaft für Forschung des Landes Nordrhein-Westfalen, Heft 193, pp. 1-91. Köln-Opladen: Westdeutscher Verlag 1969.
Sherman, M.I.: Biochemistry of differentiation of mouse trophoblast: esterase. Exp. Cell Res. 75, 449-459 (1972).
Skalko, R.G.: Methods for histologic and autoradiographic analysis of the early mouse embryo. In: Methods in Mammalian Embryology (ed. *J.C. Daniel*, jr.). pp. 238-246. San Francisco: Freeman 1971.
Spemann, H.: Experimentelle Beiträge zu einer Theorie der Entwicklung. Berlin: Springer 1936. Reprint Berlin-Heidelberg-New York: Springer 1968.

Stern, M.S.: Experimental studies on the organization of the preimplantation mouse embryo. II. Reaggregation of disaggregated embryos. J. Embryol. exp. Morph. 28, 255-261 (1972).
Stern, M.S.: Chimaeras obtained by aggregation of mouse eggs with rat eggs. Nature (Lond.) 243, 472-473 (1973).
Stern, M.S., Wilson, I.B.: Experimental studies on the organization of the preimplantation mouse embryo. I. Fusion of asynchronously cleaving eggs. J. Embryol. exp. Morph. 28, 247-254 (1972).
Tarkowski, A.K.: Experiments on the development of isolated blastomeres of mouse eggs. Nature (Lond.) 184, 1286-1287 (1959).
Tarkowski, A.K.: Mouse chimaeras developed from fused eggs. Nature (Lond.) 190, 857-860 (1961).
Tarkowski, A.K.: Embryonic and postnatal development of mouse chimeras. In: Preimplantation Stages of Pregnancy (eds. *G.E.W. Wolstenholme, M. O'Connor*), pp. 183-193. London: Churchill 1965.
Tarkowski, A.K., Wróblewska, J.: Development of blastomeres of mouse eggs isolated at the 4- and 8-cell stage. J. Embryol. exp. Morph. 18, 155-180 (1967).
Wilson, I.B., Bolton, E., Cuttler, R.H.: Preimplantation differentiation in the mouse egg as revealed by microinjection of vital markers. J. Embryol. exp. Morph. 27, 467-479 (1972).
Wimsatt, W.A.: An analysis of implantation in the bat, Myotis lucifugus lucifugus. Amer. J. Anat. 74, 355-411 (1944).
Wimsatt, W.A.: Some comparative aspects of implantation. Biol. Reprod. 12, 1-40 (1975).
Woodland, H.R., Graham, C.F.: RNA synthesis during early development of the mouse. Nature (Lond.) 221, 327-332 (1969).
Zeilmaker, G.H.: Fusion of rat and mouse morulae and formation of chimaeric blastocysts. Nature (Lond.) 242, 115-116 (1973).

Note added in Proof

After this manuscript was finished, a number of publications appeared giving additional information and supporting either theory A or B respectively. *Avendano* et al. (1975) found, in a seven-cell human embryo, two differently staining groups of blastomeres the arrangement of which resembles the polar grouping of presumptive trophoblast and embronic knot cells seen in the rabbit (see p. 70 and Fig. 1 B_2). This may support theory B. An excellent review of studies of early cell determination and differentiation using the experimental teratoma model is given by *Damjanov* and *Solter* (1974). These and additional relevant references are listed below.

Avendano, S., Croxatto, H.D., Pereda, J., Croxatto, H.B.: A seven-cell human egg recovered from the oviduct. Fertil. Steril. 26, 1167-1172 (1975).
Damjanov, I., Solter, D.: Experimental teratoma. Current Topics in Pathology (eds. *E. Grundmann, W.H. Kirsten*). 59, 69-130 (1974).
Ducibella, Th., Albertini, D.F., Anderson, E., Biggers, J.D.: The preimplantation mammalian embryo. Characterization of intercellular junctions and their appearance during development. Devel. Biol. 45, 231-250 (1975).
Ducibella, Th., Anderson, E.: Cell shape and membrane changes in the eight-cell mouse embryo. Prerequisites for morphogenesis of the blastocyst. Develop. Biol. 47, 45-58 (1975).

Ford, C.E., Evans, E.P., Gardner, R.L.: Marker chromosome analysis of two mouse chimaeras. J. Embryol. exp. Morph. 33, 447-457 (1975).

Garner, W., McLaren, A.: Cell distribution in chimaeric mouse embryos before implantation. J. Embryol. exp. Morph. 32, 495-503 (1974).

Gulyas, B.J.: A reexamination of cleavage patterns in eutherian mammalian eggs. Rotation of blastomere pairs during second cleavage in the rabbit. J. exp. Zool. 193, 235-248 (1975).

Izquierdo, L., Marticorena, P.: Alkaline phosphatase in preimplantation mouse embryos. Exp. Cell Res. 92, 399-402 (1975).

Izquierdo, L., Ortiz, M.E.: Differentiation in the mouse morulae. Wilh. Roux' Arch. 177, 67-74 (1975).

Panigel, M., Kraemer, D.C., Kalter, S.S., Smith, G.S., Heberling, R.L.: Ultrastructure of cleavage stages and preimplantation embryos of the baboon. Anat. Embryol. 147, 45-62 (1975).

Rossant, J.: Investigation of the determinative state of the mouse inner cell mass. I. Aggregation of isolated inner cell masses with morulae. J. Embryol. Exp. Morph. 33, 979-990 (1975).

Rossant, J.: Invesitgation of the determinative state of the mouse inner cell mass. II. The fate of isolated inner cell masses transferred to the oviduct. J. Embryol. exp. Morph. 33, 991-1001 (1975).

Sherman, M.I.: The role of cell-cell interaction during early mouse embryogenesis. In: The Early Development of Mammals (British Society for Developmental Biology Symposium 2), pp. 145-164. Cambridge University Press, Cambridge 1975.

Sherman, M.I.: Long term culture of cells derived from mouse blastocyts. Differentiation 3, 51-68 (1975).

III. Pharmacological and Hormonal Influences in Early Embryogenesis

The Response of the Preimplantation Embryo to Exogenous Factors

CECILIA LUTWAK-MANN

The influence of exogenous agents upon the free-lying preimplantation embryo (blastocyst) is one of the areas of inquiry that is specifically concerned with the much wider problem of maternal-embryonic relationships and exchange reactions between the embryo and its environment. The mechanisms underlying these reactions are carefully adjusted by systems which act at the level of the endometrium and its secretory products, as well as those belonging to the blastocyst, and inherent in the zona pellucida, and the highly evolved trophoblastic elements. All these processes remain throughout under strict ovarian hormonal control; hormones emanating from the blastocysts themselves are probably also involved, as it has been shown that blastocysts contain several steroid hormones (*Seamark* and *Lutwak-Mann*, 1972) and a gonadotropin type of hormone (*Haour* and *Saxena*, 1974; *Lutwak-Mann*, unpublished).

The entire problem is of major interest not only to those who are engaged in research on the morphology, physiology, biochemistry, and endocrinology of early gestation, but also to the pathologist or pharmacologist who wants to know how the early conceptus reacts to the action of drugs, hormones, deficient nutrition, exposure to radiation, maternal viremia, and a host of similar conditions.

For purely practical reasons, nearly all the experimental research work dealing with the response of the young embryo to foreign agents mediated by the maternal route has been done on small laboratory animals, chiefly the rabbit, rat, and mouse; high costs obviously preclude the use of adequate numbers of domestic animals, but some observations have been made in experiments on primates. In the past, most of the experiments have been concerned with maternally transmissible embryo-seeking agents, but currently much work is also being done under conditions in vitro.

Each laboratory species offers special advantages to the experimenter. In this respect the rabbit occupies a rather favorable position, for the following reasons. In this species ovulation occurs regularly within a reproducible time-interval (10-12 h) from the time of mating; this in turn, permits a fairly accurate assessment of embryonic age. Moreover, a fertile rabbit can be relied upon to produce, without recourse to superovulation, about 10-15 embryos. Lastly, at the preattachment stage the rabbit blastocyst reaches a relatively large size (4-5 mm diameter, 100-150 mg fresh weight), which facilitates considerably morphological or biochemical handling of this delicate and scarce material (see also *Spielmann*, this volume).

During several years of study on rabbit blastocysts (*Lutwak-Mann*, 1965, 1966, 1971, 1973; *Lutwak-Mann* and *Hay*, 1965) it became possible to accumulate information on certain features that are typical of the blastocyst stage in general, especially in the context of maternal-embryonic exchange reactions. Those characteristics that have a bearing upon pathological and pharmacological aspects of blastocyst behavior are summarized and briefly discussed below.

Blastocysts of polytocous animals show marked developmental variability. This is evident within and between litters. Such differences are, of course, equally well known in multiple human pregnancies. They are presumably partly of genetic origin and partly they may arise owing to temporal differences in the incidence of ovulations.

The existence of intra- and interlitter differences in embryonic developmental advance is probably responsible for the variability in embryonic response to drug action, a phenomenon that one commonly encounters in experiments involving maternal exposure to various agents, as well as under conditions of blastocyst exposure in vitro.

The blastocyst is a developmental stage at which there is considerable wastage under physiologic conditions. This is assessed at 10-20% in fertile laboratory animals, and nearly 50% in the human. The loss of embryos so early in pregnancy does not affect maternal health. Presumably it represents nature's way of eliminating nonviable progeny.

Contrary to opinions held a few decades ago by many obstetricians, pediatricians, and biologists, the mother's womb offers little security to the young conceptus and there is now incontrovertible evidence that embryotropic agents applied to the mother (parenterally, orally, by total body exposure) reach the free-lying blastocyst very rapidly indeed. Substances capable of penetrating into blastocysts and administered in suitable amounts and at the right time can (1) exert no damage at all, because the targets for their action are nonexistent, (2) cause a minimal to moderate embryopathic effect which may wear off in time, thus enabling the majority of these blastocysts to recover, and (3) produce severe damage, incompatible with survival; this category of agents is referred to as "embryotoxic."

The clearance rate from the blastocysts of maternally transmitted agents is usually significantly lower than their elimination from maternal peripheral body fluids, and maternal circulating levels (e.g. of an ingested drug) are not a reliable guide to drug concentrations prevailing in the fluid within the blastocyst cavity.

The delay in embryonic clearance depends to some extent upon the chemical and physical properties of the agents involved, and is due to the fact that the preimplantation embryo represents a typical low-flux compartment (*McIntosh* and *Lutwak-Mann*, 1972, 1974). Because of that, some potentially noxious drug metabolites can be expected to persist within the blastocysts, until the subsequent, more advanced, and thus more sensitive stage of organogenesis has been attained. Such a situation carries with it the risk of limb or other malformations (teratogenesis).

The blastocyst is a developmentally highly organized entity and possesses two morphologically disparate regions, namely the trophoblast and the embryo proper (referred to as embryonic disk, also inner cell mass). It has been found in experiments with a wide range of pharmacologically active substances, that there is a marked differential in the degree of responsiveness, as between the embryonic disk and the trophoblast, the former being highly vulnerable, the latter remarkably resistant to injury, incurred either by the maternal route or under conditions in vitro. This differential sensitivity denotes the existence of differences in the rate and type of metabolism in these two areas of the preimplantation embryo.

Blastocysts, in which owing to deleterious influences the disk has been largely destroyed, can nevertheless appear quite "normal" to the naked eye. Moreover, they are capable of carrying on, as purely trophoblastic vesicles, for a period of days, before finally breaking down. Therefore, when experimenting with blastocysts of any species it is mandatory to check, preferably at the outset and the termination of the experimental period, by a microscopic procedure such as e.g. the flat-mount (*Lutwak-Mann* and *Hay*, 1967), the condition of the embryos, so as to observe in detail the extent of damage, produced especially in the area of the embryo proper.

Free-lying embryos are very adaptable and able to continue growing and developing quasi-normally for limited periods of time, in extra-uterine sites and also in artificial culture media of varying composition. The remarkable resilience of the preimplantation embryo permits much interesting work to be done in vitro. It also enables a comparison to be made between effects that have been obtained in blastocysts using the same factor in experiments in vivo and in vitro.

When comparing the results of experiments in which the action of the same substance has been tested on the embryos in vivo and in vitro, one finds that whereas in some instances, both these types of experimentation yield identical results, in other cases the results differ completely. If for instance, the compound examined has been metabolized and detoxified in the maternal organism before reaching the embryos, it may become quite innocuous in vivo, but it can still prove to be embryotoxic in vitro.

The agents, of which the embryopathic effects have been tested experimentally, include various metabolites, metabolic analogues, enzyme inhibitors, hormonal deficiency or excess, nutritional influences especially in relation to vitamin lack or hypervitaminosis, exposure to ionizing radiation, alteration in partial pressure of oxygen, variation in environmental temperature, and also passage of virus.

Conditions arising from hormonal disorders, also imbalance in the composition of nutrients in the mother, act primarily upon the uterine environment, and the pathologic changes produced thereby affect the blastocysts secondarily.

Blastocysts are sometimes able to withstand the effects of moderately toxic agents. Provided the damage sustained is not extensive, the embryos can overcome it and continue to progress to term. There is of course, a danger inherent in this embryonic recuperative capacity, because even a minor loss of cells at an early stage of growth may eventually become the source of a serious defect in the individual's neonate or even late postnatal life.

Enough has probably been said to convince pathologists that the early mammalian embryo represents a most valuable and attractive object of study, provided one takes into account seriously its fundamental physiological and biochemical properties. It is self-evident that because of the embryos' inherent variability, all results of screening potentially harmful agents must be subjected to rigorous statistical appraisal (e.g., in the pharmaceutical industry). Also, seeing how much is at stake in such investigations, the responsiveness of blastocysts must be examined in more than one animal species, if one aims at clinically meaningful information.

It is conceivable that in the not too distant future fertilization of human gametes and embryonic development under conditions in vitro will both become a routine laboratory procedure. In that case, perhaps, we shall be able to do away with the need for extrapolating to man from experimental evidence obtained with animals. Progress in experiments with human embryos will, of course, depend to a large extent on agreement being reached concerning the many ethical problems involved in such research work.

References

Haour, F., Saxena, B.B.: Detection of gonadotrophin in rabbit blastocysts before implantation. Science 185, 444-445 (1974).

Lutwak-Mann, C.: Experimental embryopathy as a tool in research on animal reproduction. In: Embryopathic Activity of Drugs. Biological Council Symposium, pp. 138-151. London: Churchill 1965.

Lutwak-Mann, C.: Some physiological and biochemical properties of the mammalian blastocyst. Bull. Swiss Acad. med. Science 22, 101-122 (1966).

Lutwak-Mann, C.: The rabbit blastocyst and its environment: physiological and biochemical aspects. In: The Biology of the Blastocyst, pp. 243-260. University of Chicago Press 1971.

Lutwak-Mann, C.: Drugs and the blastocyst. In: Fetal Pharmacology, pp. 419-441. New York: Raven Press 1973.

Lutwak-Mann, C., Hay, M.F.: Maternally transmissible embryotropic agents. In: Agents Affecting Fertility. Biological Council Symposium, pp. 261-274. London: Churchill 1965.

Lutwak-Mann, C., Hay, M.F.: The blastocyst flat-mount technique in studies on embryotropic agents. In: Advances in Teratology, 2, pp. 229-238. London: Logos Press 1967.

McIntosh, J.E.A., Lutwak-Mann, C.: Zinc transport in rabbit tissues. Biochem. J. 126, 869-876 (1972).

McIntosh, J.E.A., Lutwak-Mann, C.: Calcium transport in the early conceptus and associated maternal tissues in the rabbit. Biochem. J. 138, 97-105 (1974).

Seamark, R.F., Lutwak-Mann, C.: Progestins in rabbit blastocysts. J. Reprod. Fertil. 29, 147-148 (1972).

Spielmann, H.: Embryo transfer technique and action of drugs on the preimplantation embryo. Current Topics in Pathology (eds. *E. Grundmann, W.H. Kirsten*) 62, 87-103 (1976).

Embryo Transfer Technique and Action of Drugs on the Preimplantation Embryo

H. SPIELMANN

I. Introduction

Having introduced the transplantation of mammalian preimplantation embryos as a technique to experimental teratology, *Runner* (1965) pointed out that among the offspring from transplantation experiments no increased frequency of malformations occurred and that all attempts to induce malformations by treating preimplantation embryos with drugs failed. This opinion is still shared by the majority of investigators working in experimental teratology. In a recent monograph *Wilson* (1973) concludes that there is an initial refractory period from fertilization through the cleavage and early germ-layer stages, during which little teratogenic, but appreciable lethal effects may be induced. This is followed by a period of

maximal susceptibility toward teratogenic effects that coincidences with organogenesis. *Austin* (1973) suggests that the final outcome of embryotoxic agents on the preimplantation embryo apparently depends on the number of cells killed or seriously affected: above a certain proportion, the embryo dies, below that figure, the remaining cells multiply to replace those lost and subsequent development is essentially normal.
Despite these discouraging reports I found evidence from literature and my own work for an abnormal development of mammalian embryos treated during the preimplantation period. As this problem does not seem to be solved conclusively I started this survey focusing on three main questions: (1) Do teratogenic agents reach the preimplantation embryo? (2) What are the pathogenetic mechanisms that interfere with the development of affected preimplantation embryos? (3) What is the fate of such embryos during later stages of development?

Research in this area has recently been facilitated by new methods for the in vitro culture and transplantation of preimplantation embryos of several mammalian species. The contribution of these techniques to embryonic pharmacology will be outlined. The day on which sperm is found in vaginal smears after an overnight mating period will be called day 0 p.c. (of pregnancy) in this article.

II. Toxic Effects Induced in the Preimplantation Embryo After Treatment in vivo

In this section experiments will be reviewed in which the effect of agents on the preimplantation embryo was judged by morphologic criteria usually after application of lethal or highly toxic doses.

1. Effects of Chemical Agents on Embryonic Development

Lutwak-Mann et al. (1962, 1965, 1973, this volume; *Moog* and *Lutwak-Mann*, 1958) introduced a technique for the examination of preimplantation rabbit embryos after maternal treatment on day 4 and 5 p.c. The free-lying blastocysts are removed from the uterus and under a stereoscopic microscope "flat-mounts" are prepared. The development of the rabbit blastocyst at day 5-7 p.c. can be evaluated from the morphologic features of a flat-mount preparation. The agents tested in this system included hormones, colchicine derivatives, alkylating compounds, antimetabolites, inhibitors of carbonic anhydrase, and substances with teratogenic, hepatotoxic, radiation-protective, and hypoglycemia-inducing properties (*Lutwak-Mann*, 1973). General findings emerging from these studies were the speed with which exogenous agents penetrated into the embryos even before uterine attachment and differential drug sensitivity of different areas of the rabbit blastocyst, as between the embryonic disk and the trophoblast.

Gottschewski (1963, 1964), and *Gottschewski* and *Zimmermann* (1963) used a different approach. After treatment during the blastocyst stage, they inspected rabbit embryos during organogenesis. Following the application of cyclophosphamide (40 mg/kg i.v.), nitrogen mustard (0.1 mg/kg i.v.), and thalidomide (10 mg/kg i.v.) to pregnant rabbits before implantation, they observed a high percentage of deformed fetuses on days 11, 17, and 30 p.c. They, therefore, concluded that the substances penetrate into the blastocysts even before implantation and interfere with normal development. They suggested to name abnormalities induced by this treatment "blastopathies." *Brock* and *v. Kreybig* (1964) extended *Gottschewski*'s studies to the rat. Cyclophosphamide, chlorpromazine, and thalidomide were given on day 3 p.c. and the uteri were examined for malformed and resorbed embryos on day 12. Under experimental conditions the number of implantations was not reduced when compared to controls, indicating that the blastocysts did not degenerate before implantation. The number of resorbed and malformed embryos was significantly increased after 40 mg/kg cyclophosphamide. In addition, the malformed embryos were considerably smaller than controls. On day 15 p.c. all malformed embryos were resorbed and the investigators could either only identify resorptions or normal embryos. This observation stresses the importance of the time of searching for malformations. *Gebhard* (1970) injected mice once with 10 or 20 mg/kg cyclophosphamide on day 3 p.c. He found an increased resorption rate of 30% but no malformed embryos at the end of gestation. He did not search for malformations during organogenesis. We performed similar experiments in rats with 60 mg/kg cyclophosphamide on day 3 of gestation (*Spielmann* et al., 1974). Among 30 litters on day 11 we never observed severe malformations of the brain or heart as described by *Brock* and *v. Kreybig* (1964) (Fig. 1). The embryos seemed to be retarded in development rather than grossly malformed. As the abnormal development of these embryos can further be elucidated only with the embryo transplantation technique, I will continue the discussion of this problem in Chapter IV, 2a.

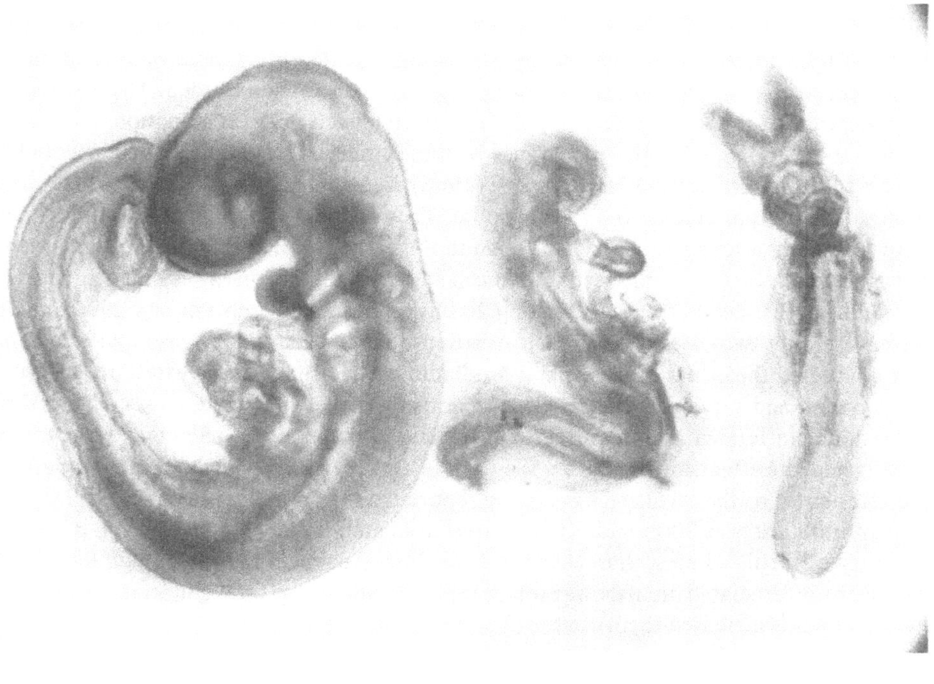

Fig. 1. Rat embryos during organogenesis (day 11) after cyclophosphamide treatment during preimplantation period. Embryo on left side is normal control, mothers of two severely retarded embryos received 60 mg/kg cyclophosphamide s.c. on day 3 p.c.

Actinomycin D has been tested during the preimplantation period in both the rat and mouse. *Wilson* (1965) found that a single dose of 300 μg/kg actinomycin D given prior to implantation, especially on day 4, prevents the implantation of blastocysts. On day 20 p.c. he found implantation sites in only 3 out of 11 females but no abnormal embryos. He did not search for malformations earlier during pregnancy. In the mouse, actinomycin D at a dose of 15 μg per animal given on day 4 p.c. interrupted pregnancy in all animals tested (*Finn* and *Martin*, 1972), these actinomycin D doses are close to the maternal LD_{50}.

2. Effects of X-Irradiation on Embryonic Development

Extensive studies on the effects of x-irradiation on hundreds of preimplantation embryos were made by *Russel* in the mouse (1950, 1956). According to these studies 200 r given before implantation (0.5-4.5 days p.c.) increase prenatal mortality in the early stages, but 100% of the embryos surviving until term appear normal. After an application of 200 r during organogenesis, 100% of the embryos are malformed or resorbed, but less than 10% are malformed when the embryos are treated with the same dose shortly before term. There are reports on malformations following x-irradiation of the preimplantation embryo. *Rugh* and *Grupp* (1959) found 2-5% exencephaly at term in 1100 mouse embryos after a single exposure to x-rays at any time from 0.5 to 9.0 days p.c. and 0% in untreated controls. They also described an unusually high susceptibility of the fertilized egg to irradiation prior to the first cleavage with 65% being resorbed after 200 r. *Inman* and *Markivee* (1963) irradiated rabbit blastocysts with 150 r and 200 r on day 3.5 p.c. and compared treated embryos with untreated littermates on day 9.5. In each treated litter there were some embryos of normal size, many stunted, and a few grossly malformed and retarded. The investigators emphasized that the severe effects would have been missed in an examination made near the end of gestation, since the appearance of these embryos suggested that they would not survive till term. *Russel* and *Montgomery* (1966) studied embryonic mortality after irradiation of the mother on days 0-4 in the mouse. They found a very high sensitivity of the embryo shortly after sperm entry and again in the early pronuclear stage. On day 1, sensitivity is relatively low in the resting two-cell stage, it then becomes very high at the beginning of the second cleavage. Among 263 living irradiated embryos and 299 controls there were only two cases of skull malformations, one in each group. *Brent* and *Bolden* (1968) studied the effect of maternal x-irradiation on the first day of gestation on embryonic development and survival in the rat. This treatment neither increased gross congenital malformations nor fetal growth retardation at the end of pregnancy. *Gibbons* and *Chang* (1973) investigated the effects of irradiating the uteri of rats on embryonic development before and after entry of the fertilized eggs. They reported a higher incidence of pre- and post-implantation losses at 300 r and 600 r in uteri irradiated on day 4 and 5, which is after the entry of fertilized eggs, than with the same dose at earlier times. Although fetal development in irradiated animals was retarded, the incidence of fetal abnormalities on days 8 and 17 was not increased significantly by a treatment before or during nidation.

These studies on x-irradiation of preimplantation embryos in vivo do not indicate whether the abnormal embryonic development of treated embryos is induced by effects directly on the embryo or primarily on the mother.

3. Passage of Agents into Preimplantation Embryos in vivo

The evidence for a penetration of chemical agents into preimplantation embryos was first derived indirectly from toxicologic data (Chapter II, 1). Radioactive labeled amino acids were the first compounds which after application to the mother could be identified in preimplantation embryos. *Edwards* and *Sirlin* (1956) instilled ^{14}C-glycine directly into the uteri of mice at the time of mating. The animals were killed at intervals and the embryos were examined for radioactivity. Blastocysts were more heavily labeled than pronuclear and two-cell embryos. *Greenwald* and *Everett* (1959) reexamined this problem by giving single injections of ^{35}S-methionine to pregnant mice. They, too, found that the blastocysts incorporated more labeled amino acid than did tubular eggs. *Lutwak-Mann*'s group (*Lutwak-Mann* et al., 1960) showed the passage of $^{32}PO_4$, ^{42}K, $^{35}SO_4$, ^{24}Na, and ^{131}J into endometrial fluid and preimplantation blastocysts of rabbits receiving these ions parenterally.

The uptake of proteins, bacteria and viruses before and during implantation was studied by *Gottschewski* and *Zimmermann* (1961, 1963) and *Zimmermann* et al. (1963). Fluorescein labeled Coxsackie virus A 9 could always be found in the uterine fluid after i.v. injection of the mother but it did not penetrate into the blastocyst fluid before attachment of the embryo to the uterine wall. Proteins and bacteria were transferred into the embryo 24 h later when the trophoblast cells have already started invading the uterine epithelium. In contrast, *Glass* (1963) was able to demonstrate the transfer of serum antigens from the maternal blood into oviductal and uterine mouse eggs with the fluorescent antibody technique.

Fabro and coworkers have extensively published data on the passage of foreign chemicals into the preimplantation rabbit embryo (*Fabro* et al., 1964, 1965, 1967; *Sieber* and *Fabro*, 1971; *Fabro*, 1973). They demonstrated an inverse relationship between the molecular weight of a radioactive compound and its uterine fluid/plasma radioactivity ratio. They also found that pregnancy can modifiy the degree to which drugs pass into the uterine fluid. In most of the compounds tested the amount of radioactivity in the uterine secretion of 6-day pregnant rabbits was not significantly different from nonpregnant animals. Following the administration of ^{3}H-nicotine, however, the radioactivity was ten times higher in the uterine fluid of pregnant animals than in nonpregnant animals. After reaching the uterine fluid the compounds penetrate the preimplantation blastocyst, where they reach different concentration levels: ^{3}H-nicotine attained a concentration in blastocysts approximately four times higher than in maternal plasma. For other compounds, e.g., caffeine, barbital, thiopental and isoniazid, the concentration in the blastocyst was approximately equal to that in the maternal plasma. The metabolites of compounds which could be identified in the preimplantation blastocyst were of maternal origin and the embryos did not seem to be able to form them. Thalidomide, e.g., undergoes spontanous hydrolysis at pH levels above 6, it is, therefore, more rapidly broken down in the more alkaline blastocelic fluid than in the uterine fluid resulting in an increase of its metabolic products in the preimplantation blastocyst (*Keberle* et al., 1966; *Fabro* et al., 1967). When using the implantation rate and volume of the conceptus as well as the outcome of pregnancy at 28 days the same investigators could not find any signs of teratogenic actions with the compounds tested for their penetration into the blastocyst.

Some of the results on the passage of substances into the rabbit blastocyst may be species specific, e.g., thalidomide. Since there are no data on the penetration of drugs into the much smaller rat and mouse blastocysts, further investigations are required in these two species.

III. Toxicology and Pharmacology of in vitro Cultured Preimplantation Embryos

During the last decade methods to support the cleavage of mammalian embryos in vitro have been improved considerably (*Brinster*, 1963; *Whitten* and *Biggers*, 1968; *Brinster*, 1973). In some species – e.g. mice and rabbits – almost all the one- or two-cell eggs placed in culture will develop into morulae or blastocysts. In the rat, however, no two-cell egg reaches the blastulae stage. When the culture requirements had been established, investigations on the effects of different metabolic inhibitors on the in vitro development have begun to elucidate characteristics of metabolic pathways of the preimplantation embryos.

1. Effects of Agents on in vitro Development

Mintz (1964) tested the specificity of actinomycin D inhibition on the mouse embryo in vitro in an autoradiographic study. The inhibitor was found to interfere primarily with nucleolar RNA synthesis, other RNA species and proteins were synthesized at a slightly re-reduced rate. *Thomson* and *Biggers* (1966) determined the dose response curves for two-cell and eight-cell mouse embryos cultured in mitomycin, actinomycin D, puromycin, and fluorophenylalanine. These studies indicated that the synthesis of some species of RNA and proteins is required for the normal progression of cleavage. Later *Thomson* (1967) extended these investigations to inhibitors of carbohydrate and energy metabolism.

The susceptibility of the early mouse embryo to actinomycin D in vitro suggests, according to *Skalko* and *Morse* (1969), that this drug affects blastocyst formation as well as blastocyst differentiation. *Monesi* et al. (1970) and *Monesi* and *Molinaro* (1971) cultured mouse embryos from the two-cell to the blastocyst stage in the presence of actinomycin D or puromycin and the labeled precursors ^{3}H-uridine and ^{3}H-leucine. This study indicated that in the mouse embryo protein synthesis and normal development depend on a continuous synthesis of RNA. A similar result was obtained by *Tasca* and *Hillman* (1970) who used the same labeled precursors and actinomycin D and cycloheximide as inhibitors.

More recently *Piko* and *Chase* (1973) cultured two-cell mouse embryos in vitro at concentrations of ethidium bromide and chloramphenicol, which inhibit mitochondrial RNA and protein synthesis exclusively, yet normal development and cellular differentiation occurred. The latter, therefore, do not seem to depend on mitochondrial genetic activity in preimplantation mouse embryos.

Some of the substances used in the in vitro studies never reach inhibitory concentrations in vivo and several embryotoxic agents are activated by maternal enzymes e.g. cyclophosphamide. In vitro inhibition studies, therefore, have to combined with embryo transfer experiments after treatment in vivo and in vitro.

2. Passage of Chemical Agents into the Preimplantation Embryo in vitro

When culturing rabbit blastocysts in vitro in the presence of tritiated water and glutethimide, *Keberle* et al. (1966) found approximately the same concentrations in the blastocoele and in the culture medium. *Sieber* and *Fabro* (1971) studied the passage of radioactive compounds into the rabbit blastocyst in vitro to obtain information on the mechanism influencing this transfer. They could show that dextran of high molecular weight does not penetrate the 6 day blastocyst, whereas 12 other compounds of smaller molecular weight, including low molecular weight dextran, salicylate, sulfanilamide, antipyrine and hexamethonium, enter the blastocyst at a rate which depends on lipid solubility, degree of ionization, and molecular weight.

IV. Transplantation of Preimplantation Embryos in Developmental Pharmacology

As the described methods for evaluating the effects of teratogenic agents on the preimplantation embryo only offer limited information, several investigators introduced the embryo transfer technique to developmental pharmacology.

1. Techniques of Embryo Transplantation

a) Transplantation of Mouse and Rat Embryos

In a series of surgical transfer experiments using pseudopregnant recipients *Noyes* and *Dickmann* established the optimal relationship of embryonic age to endometrial development for a successful transplantation in the rat (*Noyes* and *Dickmann*, 1960, 1961; *Dickmann* and *Noyes*, 1960, 1961; *Noyes* et al., 1961a) and in the mouse (*Noyes* et al., 1961b; *Doyle* et al., 1963), where they extended *McLaren* and *Michie*'s studies on the transfer of blastocysts to normal pregnant recipients (1956). The basic conclusions are similar in both species: embryos of the same age or one day older than the stage of the uterus are more likely to survive than ova 1 day younger. The simplest and most successful combination is the transplantation of blastocysts into the uterus. This approach has, therefore, been used by most investigators. As compiled in Table 1, a rate of 45-60% living term fetuses has been found in pregnant recipients if blastocysts are transferred to uteri which are synchronous or one day less advanced in development.

When transferring varying numbers of mouse embryos (2-12) on day 3 to the uteri of pseudopregnant recipients on day 2, *McLaren* (1970) found no effect in the number of embryos transferred on the rate of implantation. Smaller litters had a better survival rate after implantation. *Gosden* (1974) also reports a surprisingly high failure of implanted mouse embryos to survive to term (50%). He usually transplanted 6 embryos.

Table 1. Success rate of blastocyst transplantation to the uterus by different investigators

A Species	B Stage of donor (days p.c.).	C Stage of recipient (days p.c.)	D No. of blastocysts transferred	E No. of recipients	F Recipients showing implantations as % of E	G Viable term fetuses as % blastocysts transferred to F	H Investigator
1. Mouse	2	2	99	10		11	*McLaren* and *Michie* (1956)
2. Mouse	3	2	147	16		23	*McLaren* and *Michie*(1956)
3. Mouse	2	2	260	56	84	45	*Noyes* et al. (1961b)
4. Mouse	3	2	287	53	83	48	*Noyes* et al. (1961b)
5. Mouse	3	2	98	9	66	58	*McLaren* (1969)
6. Mouse	3	2	521	79		50	*Marsk* and *Larsson* (1974)
7. Mouse*		2	470	48	95	49	*Mullen* and *Carter* (1973)
8. Mouse*		2	125	25	69	32	*Fisher* and *Smithberg* (1974)
9. Rat	3	3	68	9	89	69	*Noyes* and *Dickmann* (1960)
10. Rat	4	3	62	9	89	66	*Noyes* and *Dickmann* (1960)
11. Rat	4	3	640	66	50	25	*Noyes* et al. (1961a)
12. Rat	4	4	324	79	87	50	*Spielmann* et al. (1974)

All recipients pseudopregnant except No. 1, 2, 6 in which G is calculated as % of E since alien and own resorptions (F) cannot be identified genetically. Transfer to both uterine horns in No. 1, 2, 3, 4, 7, 9, 10, 11; all transfers surgical except No. 6.
* Blastocysts developed from 4- and 8-cell embryos during in vitro culture for 24 h (No. 7) and 48 h (No. 8).

The transfer of embryos of different developmental age to the mouse oviduct on day O of pseudopregnancy was performed by *Bronson* and *McLaren* (1970). Regardless of the age of the embryos, between 50 and 90% of the recipients became pregnant, carrying 70% of the embryos as life fetuses to term. Despite this high success rate this method is more difficult and time consuming than the uterine transplantation and not used in routine transplantation experiments. Although it is obvious that a nonsurgical technique for transferring embryos into the uterus through the vagina might be preferable to a surgical one, this has only rarely been attempted. *Beatty* (1951) and *Tarkowski* (1959) had very poor success rates of less than 10% live term fetuses. Recently, *Marsk* and *Larsson* (1974) successfully improved the method. From 521 blastocysts transferred on day 3 through the cervix into the uteri of 2 day normal pregnant recipients they recovered 50% as living embryos, which were identified by genetic markers. In a report on the nonsurgical transfer of eggs through the cervix in rats, 25-30% of the embryos survived to term in synchronous transfer on day 4, the success rate in surgical transfers by the same investigators was 50% (*Vickery* et al., 1969).

b) Transplantation Following in vitro Culture

Since media for the culture of mouse and rabbit embryos are well-defined, attempts of transplanting in vitro developed embryos back to foster mothers have usually been performed in these two species. *McLaren* and *Biggers* (1958) cultured 8-16 cell mouse embryos for 2 days, 87% of the embryos developed into blastocysts. These blastocysts were transplanted into recipients 2 days p.c. The yield of live fetuses from cultured embryos (20%) was not significantly different from the yield which was obtained with noncultured blastocysts (21%). *Mullen* and *Carter* (1973) cultured eight-cell mouse embryos overnight and transplanted them to one or both uterine horns of 2-day pseudopregnant hosts (Table 1). Transplantation to both horns yielded a higher percentage of developing fetuses and fewer resorptions than transplantation to only one horn, the success rate was higher in hybrid than in inbred recipients. *Fisher* and *Smithberg* (1973) cultured four-cell embryos for 48 h and transplanted the blastocysts into the uterine of 1-4 day pseudopregnant hosts. The transplanted 4-day blastocysts implanted and developed into fetuses most successfully in 2-day pseudopregnant hosts, the yield of term fetuses was 32% (Table 1). From the low implantation and survival rate of transplants to 4- and 3-day hosts it is apparent that a critical period of adjustment is required before implantation of the cultured embryos can proceed.

Staples (1967) cultured rabbit blastocysts in vitro for 8, 16, and 24 h and transplanted them to hosts. The percentage of viable fetuses developing after culture for 8 h was equal to that from noncultured blastocysts (40%). *Mills* et al. (1973) performed in vitro fertilization of rabbit ova and cultured the fertilized embryos for 24 h prior to transfer at the four-cell stage. Of the embryos, 18.5% developed into term fetuses compared to 31% of normal control four-cell embryos. There is no indication for an increased rate of abnormalities among the offspring from embryos cultured in vitro during the preimplantation period and transferred to foster mothers. In addition, the fetuses from preimplantation mouse embryos on which different kinds of microsurgery had been performed (*Gardner*, 1971 and this symposium) or which had been frozen to –196°C and – 269°C for up to 1 year (*Whittingham* et al., 1972) never showed an increased malformation rate at term.

2. Transplantation of Preimplantation Embryos After Treatment in vivo or in vitro

It is possible with this method to differentiate maternal effects of an agent following treatment during the preimplantation period from direct effects on the preimplantation embryo.

a) Transplantation of Preimplantation Embryos Treated in vivo

Adams et al. (1961) used this teratologic approach first during their studies on the action of various agents on the preimplantation rabbit embryo. They recovered embryos 24 h after mating from the oviducts and transferred them for 4-6 h into the oviducts of rabbits serving as "temporary recipients" which had received injections of either mercaptopurine or azaguanine (110-180 mg/kg). After this "in vivo treatment period" the embryos were transferred to one uterine horn of a pseudopregnant recipient, untreated controls to the other horn. One cleavage division usually occurred during the in vivo treatment. At autopsy between days 9-14 p.c. the number of implanted and viable embryos was reduced when compared with the controls. It was concluded that the effect of the purine analogues became manifest about the time of or after implantation. *Glass* and *Lin* (1963) studied the development of x-irradiated an nonirradiated recipient females. The cross transfer of irradiation and nonirradiated oocytes to irradiated and nonirradiated hosts allowed to distinguish between effects of treatment on the oocytes and on the mother (uterus). Their results indicate that nonirradiated donor eggs were more than twice as successful as eggs treated with 250 r in implantation rate and development.

Finn and coworkers investigated the inhibition of the implantation reaction in the mouse by actinomycin D (*Finn* and *Martin*, 1972; *Finn* and *Bredl*, 1973; *Pollard* et al., 1973). From electron micrograph studies it appeared that the blastocyst is stimulated to develop in treated mothers, but decidualization and the degeneration of the uterine epithelium around the blastocyst is inhibited. The blastocyst transfer technique was used to distinguish the effect of the drug on the uterus from effects on the blastocyst. Three treatment groups were investigated. In the first neither donors nor pseudopregnant recipients received actinomycin D, in the second the donor animals were treated 4 h before transfer while the recipients were untreated, and in the final group the recipients received the drug 4 h before the transfer of untreated blastocysts. At the examination 24 h and 48 h after transfer, the treated blastocysts in untreated recipients did not differ from controls. When untreated blastocysts were transplanted to treated recipients the uterine reaction was the same as in animals treated without any transplantation. This suggests clearly that actinomycin D in these experiments acts on the uterus and has no effect on the blastocyst.

In similar approach we tried to find out if the effects observed after treatment of pregnant rats with cyclophosphamide during the preimplantation period are due to the direct effect of the drug on the embryo or if the effect is primarily due to an interference with the maternal organism (*Spielmann* et al., 1974). We transferred blastocysts 24 h after maternal treatment to untreated pseudopregnant foster mothers. The transplantations were performed synchronously on day 4 and varying concentrations of cyclophosphamide and actinomycin D were used. As demonstrated in Figure 2, with increasing doses of cyclophosphamide the number of implantations and living fetuses at term decreased. In the actinomycin D group doses even higher than the maternal LD_{50} (donor) hardly affected the survival of the embryos after transfer. It is possible to use such high doses in short-term experiments because of the delayed toxicity. Our interpretation of the data is that during the 24 h treatment in vivo cyclophosphamide acts primarily on the embryo, actinomycin D on the other hand, in agreement with *Finn*'s results (*Finn* et al., 1972, 1973), predominantly on the uterus. At term in both experimental series, the malformation rate among the transferred treated embryos was not increased in the small number examined.

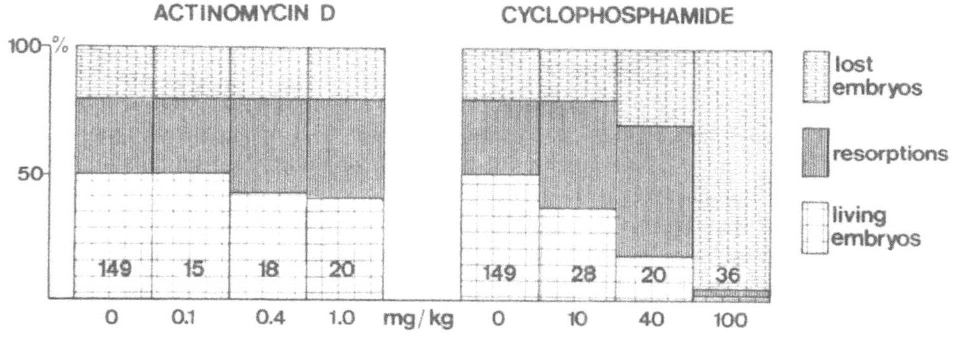

Fig. 2. Examination of fetuses at term after transplantation of preimplantation rat embryos 24 h after maternal treatment on day 3 to pseudopregnant host on day 4 p.c. Total number of embryos transplanted is indicated by numbers in columns, these numbers also serve as 100% values. Maternal LD_{50}: 180 mg/kg s.c. for cyclophosphamide and 0.5 mg/kg i.p. for actinomycin D

b) Transplantation of Preimplantation Embryos Treated During the Culture Period

Fisher (1972) studied the effect of x-irradiation on in vitro development of mouse preimplantation embryos as well as on in vivo development after transplantation. Of the four-cell embryos, 50% developed into blastocysts following 300 r in vitro irradiation and after 600 r in vivo irradiation of the mother. When transplanting the blastocysts that had developed after 48 h in culture, no fetal development occurred for embryos irradiated in vitro with more than 170 r and with more than 380 r in vivo. It is concluded that twice as much in vivo irradiation is required to equal in vitro lethal effects for the preimplantation mouse embryo. No malformations were found.

In addition, *Fisher* and *Smithberg* (1972) investigated the effects of in vitro exposure of preimplantation mouse embryos to different concentrations of trypan blue. Four-cell embryos were cultured for 48 h, the percentage of developing blastocysts was recorded and some of the blastocysts were transferred to recipients on day 2 of psuedopregnancy. Significantly fewer fetuses developed after 48 h exposure to concentrations of 0.08% trypan blue, and the percentage of resorptions increased with increasing concentrations. The only abnormality found in treated fetuses was subdermal hematoma, which is also produced by maternal administration of trypan blue after implantation (*Beck* and *Lloyd*, 1966). *Lin* and *Monie* (1973) performed identical studies. They also injected trypan blue into the blastocoele of embryos before transplantation. In the group of embryos cultured in the presence of trypan blue as well as in the group that received the injections, there was one embryo at term which displayed a large hematoma on the anterior thoracic wall. Both groups of investigators conclude that despite these minor abnormalities, the drug has an all-or-none effect on the preimplantation embryo.

Bell and *Glass* (1975) cultured blastocysts for 2 h in the presence of 10^{-5}-10^{-1} μg/ml Actinomycin D in vitro and transferred these blastocysts to pseudopregnant recipients. 15 days later they observed a dose dependant reduction in both the rate of implantation and the fetal weight in concentrations between 10^{-3} to 10^{-1} μg/ml Actinomycin D. This result and the absence of malformations among the surviving fetuses confirms for these investigators that the preimplantation embryo is relatively impervious to teratogenic agents.

Piko and *Chase* (1973) cultured preimplantation mouse embryos in the presence of different ethidium bromide chloramphenicol concentrations for 48 h. When transferring blastocysts that had developed at concentrations of the drugs, which inhibit mitochondrial RNA- and protein synthesis exclusively, into pseudopregnant recipients, no abnormalities were found in the 15 live term fetuses that had developed. *Snow* (1973) cultured mouse embryos from the two-cell stage to blastocysts in the presence of ^{3}H-thymidine at different concentrations. Selective cell death occurred at the 16-cell stage, the cells which were most susceptible to ^{3}H-Tdr damage were those normally contributing to the inner cell mass. At higher concentrations, blastocysts with no inner cell mass, but entirely consisting of trophoblast developed. These embryos were capable of further differentiation after transplantation to recipients, but only as invasive trophoblast cells with no fetal tissue.

The number of treated transplanted embryos examined for malformations at term is rather small in most studies. If no malformation is found, e.g., among 40 fetuses, there is still a 10% chance of malformations occurring (*Sachs*, 1969). Additional transplantation experiments, therefore, have to be performed to get a reliable statistic on the malformation rate among the offspring from treated transplanted embryos.

3. The Embryo Transfer Technique in Developmental Pharmacology After the Preimplantation Period

To elucidate uterine genetic factors modifying the effect of teratogenic agents after implantation *Marsk* et al. (1971) used the embryo transfer technique. After the injection of 62.5 mg/kg cortisone on days 11-14 p.c., cleft palate is induced in 100% of the embryos of A/jax mice and in 12-20% of CBA mice. When reciprocal crosses showed that fetuses growing in an A/Jax mother developed a higher percentage of cleft palate than in a CBA mother and an x-linked inheritance could be excluded, the uterine influence was further investigated. Homozygote CBA blastocysts were transferred to A/Jax foster mothers and vice versa. A/Jax fetuses were not protected in the CBA uterus and CBA fetuses in A/Jax had no increase in frequency of cleft palate. The maternal influence observed in the reciprocal crosses could, therefore, not be interpreted as uterine. *Takano* et al. (1972) used the same approach and found in contrast to *Marsk* et al. (1971) a significant influence of the maternal genome (uterine environment) on cleft palate induced by cortisone when performing the transfer between the strains C57BL/10J and C 57.

V. Conclusions

Teratogenic, toxic, and lethal effects of different agents on preimplantation embryos after maternal treatment can be found only if the evaluation is performed at the period of organogenesis, since the damaged embryos rarely survive the total prenatal period up to term. These effects can reproducibly be demonstrated after cyclophosphamide treatment and may also be found following treatment with other agents if the embryos are examined more carefully. Furthermore, in several studies the number of term fetuses that developed from treated embryos is too small to exclude a 10% chance of malformations. The conclusion that after treatment during the preimplantation period all embryos die during implantation or survive to term, therefore, has to be modified. It still has to be investigated whether treated embryos that die during organogenesis are mal-malformed or retarded.

The mechanisms of the induction of teratogenic effects in the preimplantation embryo and their interference with developmental processes after implantation is still open to question. The treatment could predominantly interfere with the maternal or the embryonic metabolism. There is evidence for both possibilities. Actinomycin D inhibits the normal decidualization of the uterus, the embryos in these animals cannot implant but die. Cyclophosphamide, on the other hand, has a direct effect on the embryos as suggested by our transplantation experiments. The affected embryos survive the period of implantation but die later during development. A retarded clearance of the substances from the blastocele could also explain their effects at later stages.

Since some of the effective agents are known to be mutagenic, e.g., cyclophosphamide and x-irradiation, the malformations and retardations may be induced by somatic mutations. Studies on trisomic mouse embryos (*Gropp*, this volume) support this hypothesis since embryos carrying this chromosomal imbalance frequently die severely malformed during organogenesis. Since somatic mutations do not induce clear-cut malformation syndromes, a wide pattern of abnormalities can be expected. This makes an evaluation even more difficult today. Furthermore, most of the statements on the absence of malformations after treatment during the preimplantation period are confined to gross morphologic examinations at term and do not take into account embryonic death during organogenesis or abnormalities which manifest themselves postnatally. I am, therefore, convinced that further investigations are required to understand the action of teratogens during the earliest period of embryonic development.

Uterine influences on the susceptibility to teratogens during the whole gestational period can be elucidated with the embryo transfer technique. The direct action of teratogens on the preimplantation embryo can, therefore, be separated from effects predominantly on the uterus. Further support of the results on the preimplantation embryo can be obtained by combining in vitro culture studies with transplantation experiments.

Acknowledgment. I am most grateful to Dr. *Diether Neubert* who introduced me to fetal pharmacology and who critically discussed several aspects of this review.
This work was supported by grants from the Deutsche Forschungsgemeinschaft, given to the Sonderforschungsbereich 29, "Embryonale Entwicklung und Differenzierung".

References

Adams, C.E., Hay, M.F., Lutwak-Mann, C.: The action of various agents upon the rabbit embryo. J. Embryol. exp. Morph. 9, 468-491 (1961).

Austin, C.R.: Embryo transfer and sensitivity to teratogenesis. Nature (Lond.) 244, 333-334 (1973).

Beatty, R.A.: Transplantation of mouse eggs. Nature (Lond.) 168, 995 (1951).

Beck, F., Lloyd, J.B.: The teratogenic effects of azo dyes. In: Advances in Teratology, 1 (ed. *D.H.M. Woolam*) p. 131-193 (1966).

Bell, P.S., Glass, R.H.: Development of the mouse blastocyst after actinomycin D treatment. Fertil. and Steril. 26, 449-454 (1975).

Brent, R.L., Bolden, B.T.: Indirect effect of x-irradiation on embryonic development: utilization of high doses of maternal irradiation on the first day of gestation. Radiat. Res. 36, 563-570 (1968).

Brinster, R.L.: A method for in vitro cultivation of mouse ova from two-cell to blastocyst. Exp. Cell Res. 32, 205-208 (1963).

Brinster, R.L.: Cultivation of the mammalian embryo. In: Growth nutrition and metabolism of cell in culture, Vol. 2 (eds. *G.H. Rothblant, V.C. Cristofalo*), pp. 251-286. New York–London: Academic Press 1972.

Brock, N., von Kreybig, Th.: Experimenteller Beitrag zur Prüfung teratogener Wirkungen von Arzneimitteln an der Laboratoriumsratte. Naunyn-Schmiedeberg's Arch. exp. Path. Pharmak. 249, 117-145 (1964).

Bronson, R.A., McLaren, A.: Transfer to the mouse oviduct of eggs with and without the zona pellucida. J. Reprod. Fertil. 22, 129-137 (1970).

Dickmann, Z., Noyes, R.W.: The fate of ova transferred into the uterus of the rat. J. Reprod. Fertil. 1, 197-212 (1960).

Dickmann, Z., Noyes, R.W.: The zona pellucida at the time of implantation. Fertil. and Steril. 12, 310-318 (1961).

Doyle, L.L., Gates, A.H., Noyes, R.W.: Asynchronous transfer of mouse ova. Fertil. and Steril. 14, 215-225 (1963).

Edwards, R.G., Sirlin, J.L.: Studies in gametogenesis, fertilization and early development in the mouse, using radioactive tracers. Proc. II. Wld. Congr. Fertil. and Steril., Naples (1956).

Fabro, S.: Passage of drugs and other chemicals into the uterine fluids and preimplantation blastocyst. In: Fetal Pharmacology (ed. *L. Boréus*), pp. 443-459. New York: Raven Press 1973.

Fabro, S., Schumacher, H., Smith, R.L., Williams, R.T.: Identification of thalidomide in rabbit blastocysts. Nature (Lond.) 201, 1125-1126 (1964).

Fabro, S., Smith, R.L., Williams, R.T.: The fate of the hydrolysis products of thalidomide in the pregnant rabbit. Biochem. J. 104, 570-574 (1967).

Finn, C.A., Bredl, J.C.S.: Studies on the development of the implantation reaction in the mouse uterus: influence of actinomycin D. J. Reprod. Fertil. 34, 247-253 (1973).

Finn, C.A., Martin, L.: Temporary interruption of the morphogenesis of deciduomata in the mouse uterus by actinomycin D. J. Reprod. Fertil. 31, 353-358 (1972).

Fisher, D.L.: In-vitro and in vivo x-irradiation of the preimplantation mouse embryo. Anat. Rec. 172, 310 (abstract) (1972).

Fisher, D.L., Smithberg, M.: Early and late effects of in vitro exposure of preimplantation mouse embryos to trypan blue. Teratology 6, 159-166 (1972).

Fisher, D.L., Smithberg, M.: Host-transplant relationship of cultured mouse embryos. J. exp. Zool. 183, 263-266 (1973).

Gardner, R.L.: Maniuplations on the blastocyst. In: Advances in the Biosciences 6 (ed. *G. Raspé*), pp. 279-296. Oxford, Braunschweig: Pergamon Press, Vieweg 1971.

Gebhardt, D.D.E.: The embryo lethal and teratogenic effects of cyclophosphamide on mouse embryos. Teratology 3, 273-278 (1970).

Gibbons, A.F.E., Chang, M.C.: Indirect effects of x-irradiation on embryonic development: irradiation of the exteriorized rat uterus. Biol. Reprod. 9, 133-141 (1973).

Glass, L.E.: Transfer of native and foreign serum antigens to oviductal mouse eggs. Amer. Zool. 3, 135-156 (1963).

Glass, L.E., Lin, T.P.: Development of x-irradiated and non-irradiated mouse oocytes transplanted to x-irradiated and non-irradiated recipient females. J. cell. comp. Physiol. 61, 53-60 (1963).

Gosden, R.G.: Survival of transferred C57B mouse embryos: effects of age of donor and recipient. Fertil. and Steril. 25, 348-351 (1974).

Gottschewski, G.H.M.: Das Entstehen bestimmter äußerer Mißbildungen als Folge exogener Reize beim Säugetier. Med. Welt 50, 2545-2556 (1963).

Gottschewski, G.H.M.: Mammalian blastopathies due to drugs. Nature (Lond.) 201, 1232-1233 (1964).

Gottschewski, G.H.M., Zimmermann, W.: Nachweis von Proteinen und TMV-Antikörpern in der Dottersackflüssigkeit von Kaninchenfruchtblasen zur Zeit der Nidation. Naturwissenschaften 48, 557 (1961).

Gottschewski, G.H.M., Zimmermann, W.: Auslösung von Blastopathien beim Säugetier durch Cyclophosphamid und Thalidomid. Naturwissenschaften 50, 525 (1963).

Greenwald, G.S., Everett, N.B.: The incorporation of S^{35}-methionine by the uterus and the ova of the mouse. Anat. Rec. 143, 171-184 (1959).

Gropp, A., Kolbus, U., Giers, D.: Systematic approach to the study of trisomy in the mouse II. Cytogenet. Cell Genet. 14, 42-62 (1975).

Inman, O., Markivee, C.R.: Gross effects on rabbit embryos and membranes of x-irradiation in the blastocyst stage. Anat. Rec. 147, 139-147 (1963).

Keberle, H., Schmid, K., Faigle, J.W., Fritz, H., Loustalot, P.: Über die Penetration von körperfremden Stoffen in den jungen Wirbeltierkeim. Bull. schweiz. Akad. med. Wiss. 22, 134-152 (1966).

Lin, T.P., Monie, I.W.: The development of mouse blastocysts injected with or cultured in, trypan blue solution. J. Reprod. Fertil. 32, 149-151 (1973).

Lutwak-Mann, C.: Drugs and the blastocyst. In: Fetal Pharmacology (ed. *L. Boréus*), pp. 419-441. New York: Raven Press 1973.

Lutwak-Mann, C., Boursnell, J.C., Bennett, J.P.: Blastocyst-uterine relationships: Uptake of radioactive ions by the early rabbit embryo and its environment. J. Reprod. Fertil. 1, 169-185 (1960).

Lutwak-Mann, C., Hay, M.F.: Effect on the early embryo of agents administered to the mother. Brit. med. J. 2, 944-946 (1962).

Lutwak-Mann, C., Hay, M.F.: Effect of 2-Deoxy-D-glucose on the rabbit blastocyst. J. Reprod. Fertil. 10, 133-135 (1965).

Marsk, L., Larsson, K.S.: A simple method for non-surgical blastocyst transfer in mice. J. Reprod. Fertil. 37, 393-398 (1974).

Marsk, L., Theorell, M., Larsson, K.S.: Transfer of blastocysts as applied to experimental teratology. Nature (Lond.) 234, 358-359 (1971).

McLaren, A.: The fate of very small litters produced by egg transfer in mice. J. Endocr. 47, 87-94 (1970).

McLaren, A., Biggers, J.D.: Successful development and birth of mice cultivated in vitro as early embryos. Nature (Lond.) 182, 877-878 (1958).

McLaren, A., Michie, D.: Studies on the transfer of fertilized mouse eggs to uterine foster-mothers I. Factors affecting the implantation and survival of native and transferred eggs. J. exp. Biol. 33, 394-415 (1956).

Mills, J.A., Jeitles, C.G., Brackett, B.G.: Embryo transfer following in vitro and in vivo fertilization of rabbit ova. Fertil. and Steril. 24, 602-608 (1973).

Mintz, B.: Synthetic processes and early development in the mammalian eggs. J. exp. Zool. 157, 85-100 (1964).

Monesi, V., Molinaro, M.: Macromolecular synthesis and effect of metabolic inhibitors during preimplantation development in the mouse. In: Advances in the Biosciences 6 (ed. *G. Raspé*), pp. 101-117. Oxford, Braunschweig: Pergamon Press, Vieweg 1971.

Monesi, V., Molinaro, M., Spaletta, E., Davroli, C.: Effect of metabolic inhibitors on macromolecular synthesis and early development in the mouse embryo. Exp. Cell Res. 59, 197-206 (1970).

Mullen, R.J., Carter, S.C.: Efficiency of transplanting normal, zona-free and chimeric embryos to one and both uterine horns of inbred and hybrid mice. Biol. Reprod. 9, 111-115 (1973).

Noyes, R.W., Dickmann, Z.: Relationship of ovular age to endometrial development. J. Reprod. Fertil. 1, 186-196 (1960).

Noyes, R.W., Dickmann, Z.: Survival of ova transferred into the oviduct of the rat. Fertil. and Steril. 12, 67-79 (1961).

Noyes, R.W., Doyle, L.L., Bentley, D.L.: Effect of preimplantation development on foetal weight in the rat. J. Reprod.Fertil. 2, 238-245 (1961a).

Noyes, R.W., Doyle, L.L., Gates, A.H., Bentley, D.L.: Ovular maturation and fetal development. Fertil. and Steril. 12, 405-416 (1961b).

Pikó, L., Chase, D.G.: Role of the mitochondrial genome during early development in mice. J. Cell Biol. 58, 357-378 (1973).

Pollard, R.M., Bredl, J.C.S., Finn, C.A.: The effect of actinomycin D on the attachment reaction of implantation in mice. J. Reprod. Fertil. 33, 343-345 (1973).

Rugh, R., Grupp, E.: Exencephalia following x-irradiation of the pre-implantation mammalian embryo. J. Neuropath. exp. Neurol. 18, 468-481 (1959).

Runner, M.N.: Transplantation of mammalian ova: blastocyst and earlier stages. In: Teratology, Principles and Techniques (eds. *J.G. Wilson, J. Warkany*), pp. 104-111. Chicago-London: University of Chicago Press 1965.

Russel, L.B.: X-ray induced developmental abnormalities in the mouse and their use in the analysis of embryological patterns. I. External and gross visceral changes. J. exp. Zool. 114, 545-602 (1950).

Russel, L.B.: X-ray induced developmental abnormalities in the mouse and their use in the analysis of embryological patterns. II. Abnormalities of the vertebral column and thorax. J. exp. Zool. 131, 329-390 (1956).

Russel, L.B., Montgomery, C.S.: Radiation-sensitivity differences within cell-division cycles during mouse cleavage. Int. J. Radiat. Biol. 10, 151-164 (1966).

Sachs, L.: Statistische Auswertungsmethoden, 2nd ed. Berlin–Heidelberg–New York: Springer 1969.

Sieber, S.M., Fabro, S.: Identification of drugs in the preimplantation blastocyst and in the plasma, urine secretion and uterine of the pregnant rabbit. J. Pharmacol. exp. Ther. 176, 65-75 (1971).

Skalko, R.G., Morse, J.M.D.: The differential response of the early mouse embryo to actinomycin D treatment in vitro. Teratology 2, 47-54 (1969).

Snow, M.H.L.: Abnormal development of pre-implantation mouse embryos grown in vitro with ^{3}H-thymidine. J. Embryol. exp. Morph. 29, 601-615 (1973).

Spielmann, H., Litzenbörger, H., Lücke, I.: Transplantation of blastocysts as a tool in teratologic studies on the preimplantation period of the rat. Naunyn-Schmideberg's Arch. Pharmacol. 282, Suppl., R94 (abstract) (1974).

Staples, R.E.: Development of 5-day rabbit blastocysts after culture at 37°C. J. Reprod. Fertil. 13, 369-372 (1967).

Takano, K., Peterson, A.C., Biddle, F.G., Miller, J.R.: Analysis of cleft palate induction by by a glucocorticoid in mice: an application of the egg transfer technique in teratology. Teratology 6, 119-120 (abstract) (1972).

Tarkowski, A.K.: Experiments on the transplantation of ova in mice. Acta theriol. 2, 251-267 (1959).

Tasca, R.J., Hillman, N.: Effects of actinomycin D and cycloheximide on RNA and protein synthesis in cleavage stage mouse embryos. Nature (Lond.) 225, 1022-1025 (1970).

Thomson, J.L.: Effect of inhibitors of carbohydrate metabolism on the development of preimplantation mouse embryos. Exp. Cell Res. 46, 252-262 (1967).

Thomson, J.L., Biggers, J.D.: Effect of inhibitors of protein synthesis on the development of preimplantation mouse embryos. Exp. Cell Res. 41, 411-427 (1966).

Vickery, B.H., Erickson, G.I., Bennett, J.P.: Nonsurgical transfer of eggs through the cervix in rats. Endocrinology 85, 1202-1203 (1969).

Wilson, J.G.: Effects of acute and chronic treatment with actinomycin D on pregnancy and the fetus in the rat. Harper Hosp. Bull. 24, 109-118 (1966).

Wilson, J.G.: Environment and Birth Defects. New York–London: Academic Press 1973.

Whitten, W.K., Biggers, J.D.: Complete culture of the preimplantation stages of the mouse in vitro. J. Reprod. Fertil. 17, 399-401 (1968).

Whittingham, D.G., Leibo, S.P., Mazur, P.: Survival of mouse embryos frozen to −196° and −269°C. Science 178, 411-414 (1972).

Zimmermann, W., Gottschewski, G.H.M., Flamm, H., Kunz, Ch.: Experimentelle Untersuchungen über die Aufnahme von Eiweiß, Viren und Bakterien während der Embryogenese des Kaninchens. Develop. Biol. 6, 233-249 (1963).

Uterine Secretion Protein Patterns Under Hormonal Influences

H.M. BEIER

I. Ovarian Hormones in Estrous and Pregnancy

It is a fundamental endocrine phenomenon that the ovarian hormones control the oviducal and uterine functions, the significance of which becomes evident in sperm capacitation, egg fertilization, cleavage, blastocyst development, and implantation. Throughout all stages of pregnancy, embryologists and endocrinologists have influenced experimentally mammalian pregnancy by ovariectomy, and have tried by various means and with different success to replace the lacking ovarian hormones. This has been a useful tool in demonstrating the essential role of estrogens and progesterone to maintain pregnancy, as well as to show roughly the quantitation of these essential steroids on a daily or even shorter basis. Naturally, this experimental approach turned out insufficently in the attempt to obtain the actual steroid levels within the maternal blood stream during pregnancy. However, the recent development of highly sophisticated methods by using radioimmune reactions has presented much information about the physiological hormone concentrations, turnover rates, binding at the target tissue's cellular level, and elimination from tissues and the circulation.

The rabbit has been extensively studied as one of the model laboratory animals on which are based much information concerning the dominating role of progestins and estrogens in the regulation of pregnancy. The concentrations of progesterone, 20 α-dihydroprogestrone, estrone, and 17β-estradiol in the blood plasma of rabbits during the course of gestation have been determined by gas liquid chromatography and radioimmunoassay in several laboratories (*Eaton* and *Hilliard*, 1971; *Hasan* et al., 1971; *Shaikh* and *Harper*, 1972; *Challis* et al., 1973; *Baldwin* and *Stabenfeldt*, 1974; *Hilliard* et al., 1974). In the investigations by *Challis* et al. (1973) the highest mean concentration of progesterone in the peripheral blood plasma was observed around midpregnancy (day 15); this peak value of the peripheral plasma seems to parallel the progesterone concentration in ovarian venous plasma, as described by previous authors (*Mikhail* et al., 1961; *Hilliard* et al., 1968). Generally, it is accepted that the profile of progesterone concentrations in blood plasma during the course of pregnancy represents a monophasic peak with a gradual increase from ovulation up to days 12-15 p.c., and a continuous decrease thereafter until parturition, as depicted in Figure 1. The concentrations of estrogens in peripheral and ovarian venous plasma in the rabbit have received attention by several investigators (*Eaton* and *Hilliard*, 1971; *Hilliard* and *Eaton*, 1971; *Shaik* and *Harper*, 1972; *Hilliard*, 1973; *Hilliard* et al., 1974; *Challis* et al., 1973). The profile of estrogen concentrations throughout the whole course of pregnancy has been reported only by *Challis* et al. (1973). There is an initial peak of 17β-estradiol after ovulation, around day 6 just prior to implantation. After this preimplantation peak 17β-estradiol concentrations again increase to day 15, and are followed by a decline. Finally, there exists a prepartum rise, the magnitude of which appears to be sluggish (Fig. 2). Increasing plasma levels for estrogens during late pregnancy and prior to parturition have been described in other mammals, too (*Davies* and *Ryan*, 1972). It is also very important to keep in mind that the ovarian steroids are exerting their biological effects in combined actions, or at least by balanced actions corresponding to their sensitively balanced plasma concentrations (*Challis* et al., 1973), as shown in Figure 3.

Detailed investigations have demonstrated a typical preovulatory steroid pattern in rabbit blood plasma (*Eaton* and *Hilliard*, 1971; *Challis* et al., 1973; *Fuchs* et al., 1974). Progesterone, 20α-dihydroprogesterone, and 17β-estradiol release can be induced by a single small injection of LH (0.5 μg/kg, i.v.) or HCG (25 I.U./animal, i.v.), applied to estrous rabbits. This experimental approach mimics what appears physiologically, the immediate induction of steroid release shortly after natural mating. The highest peak is represented by the interstitial release of 20α-dihydroprogesterone (*Okano* et al., 1966; *Fuchs* et al., 1974), which is followed directly by progesterone. Maximal values are reached as early as 30 min after mating or after the gonadotropin injection. The levels gradually decline to estrous values by the end of day 2 and on day 3. It is a characteristic feature in the rabbit, that the plasma concentrations of all ovarian steroids are approximately down to estrous levels around the time of ovulation, which usually occurs 10 h after mating (p.c.). It may be pointed out here, that the rabbit is a reflex ovulator and needs a mating stimulus or an exogenous LH surge for ovulation.

With particular attention to the different reproductive situations as normal pregnancy and pseudopregnancy, it is interesting to refer to the investigations of *Fuchs* et al. (1974). These authors found no differences in steroid release before and after ovulation up to day 3 p.c. in female rabbits, which were mated with intact or with vasectomized males, giving rise to normal pregnancy or pseudopregnancy, respectively. In normal pregnant animals plasma progesterone and 20α-dihydroprogesterone start their typical postovulatory increase on day 4, followed by a more rapid increase on subsequent days 5 and 6. By contrast, in pseudopregnant animals the increased level in plasma progesterone appears with considerable delay and does not reach the level of pregnant rabbits. On day 6, the progesterone concentration in plasma of normal pregnancy is significantly higher (12.44 ± 1.78 ng/ml) than in pseudopregnancy (4.32 ± 1.18 ng/ml). Generally, there are great variations in the individual levels of the steroids reported for rabbits and other mammals.

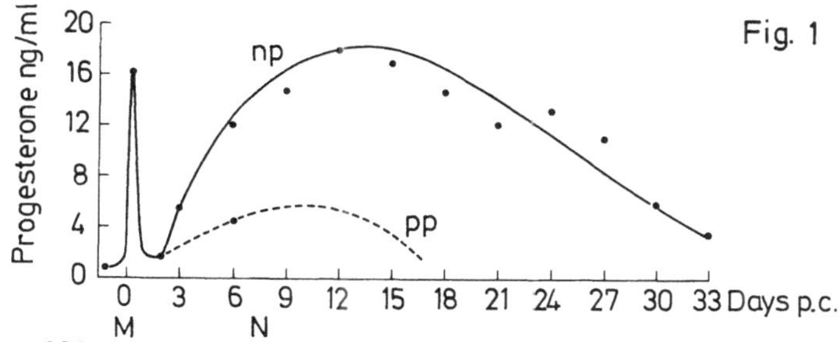

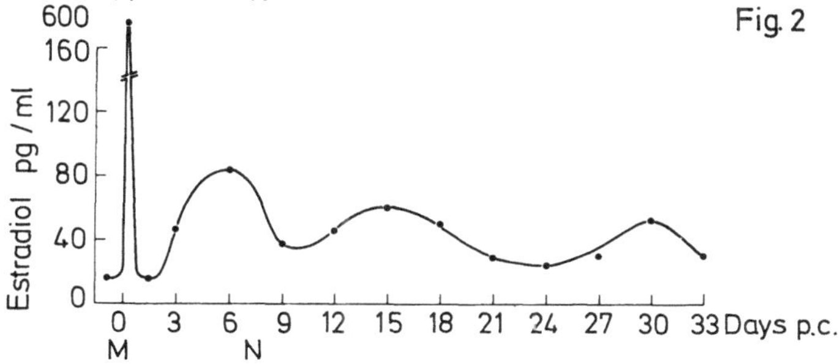

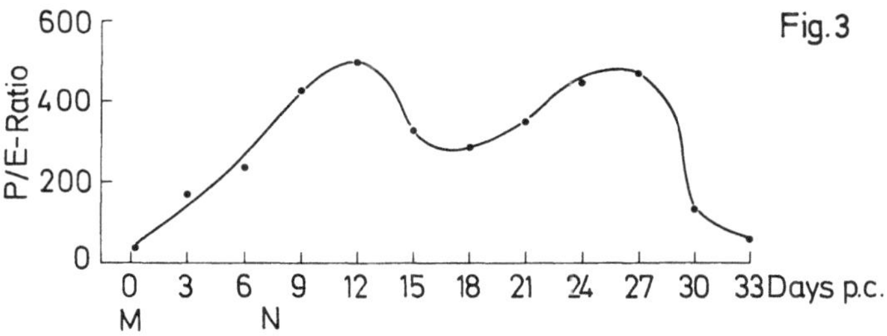

Fig. 1. Peripheral plasma progesterone concentrations in rabbits throughout pregnancy (np, normal pregnancy; pp, pseudopregnancy). M = day of mating, which is commonly designated as day 0; N = day of implantation, P = day of parturition. Data are compiled from literature (*Hasan* et al., 1971; *Hilliard* et al., 1971; *Challis* et al., 1973; *Fuchs* et al., 1974)

Fig. 2. Peripheral plasma 17β-estradiol concentrations in rabbits throughout pregnancy. M = day of mating; N = day of implantation; P = day of parturition. As for progesterone shown in Figure 1, peripheral plasma estrogen concentrations during time of ovulation (10 h and during following time of gamete transport via oviduct (day 1–day 2) are remarkably low. Data are compiled from literature (*Hilliard* et al., 1971; *Challis* et al., 1973)

Fig. 3. Ratio of progesterone to 17β-estradiol in peripheral plasma of rabbits throughout pregnancy, as demonstrated in Figures 1-2. Diagram modified after *Challis* et al. (1973)

Probably there exists an influence on the absolute values by the number of embryos, as it has also been determined in pregnant women with twins (*Hasan* et al., 1971).

II. Withdrawal of Ovarian Hormones by Ovariectomy

The normal pregnancy in the rabbit can only be maintained when enough corpora lutea exist to produce a sufficient amount of progestins. If all corpora lutea disappear by experimental ovariectomy or enucleation, all embryos must degenerate. As shown by *Lutwak-Mann* et al. (1962), total ovariectomy in the rabbit performed on day 6 p.c. inhibits implantation. In the course of these experiments *Lutwak-Mann* and coworkers found that the embryoblast keeps its mitotic activity significantly longer after hormone withdrawal than the cells of the trophoblast. Replacement of progesterone is effective in maintance of implantation and pregnancy. The rabbit blastocyst obviously does not need an preimplantational estrogen surge, like the early mouse and rat embryo. Since estrogens, 17β-estradiol in particular, turned out to be essential luteotropic factors in the rabbit, it is conceivable that only progesterone substitution is needed after ovariectomy. From ovulation up to implantation a daily dose of 1 mg of progesterone is effective, whereas after implantation, a dose of 2 mg/day is beneficial (*Hafez* and *Pincus*, 1956; *Hafez*, 1964).

III. Hormone-Sensitive Protein Secretion in the Genital Tract

Analysis of the composition of genital tract fluids at different stages of estrous and preimplantation yields a series of apparently typical time-specific biochemical patterns. If the secretory products of the different genital tract compartments which are passed by the essential reproductive requisites, oocyte and spermatozoa, are compared with regard to obvious time-specific biochemical changes, the uterine secretion by far shows the most interesting features in comparison with follicular fluid, oviductal, cervical, and vaginal secretion. Both the maternal and the embryonic systems require a strict synchronization in terms of developmentally defined time and substrate sequences, this is particularly true within the uterus. In the rabbit, the blastocyst has entered the uterine lumen early at day 4 post coitum. Until implantation on day 7 and 8, this blastocyst grows with a considerable increase in its diameter from 0.5 mm to 6.0 or even 7.5 mm.

Uterine secretion in the rabbit contains several remarkable components, which have been analyzed as specific secretory proteins showing a hormonal control of their synthesis and secretion (Fig. 4). During the time of free blastocyst movement within the uterine lumen, the endometrial epithelia deliver – among several other biochemical components – nine different proteins, which have been isolated, described, and identified during recent years, mainly in the author's laboratory (*Beier*, 1967, 1968a, b, 1970, 1973, 1974a). Among these (see Table 1), we find as uterine specific secretions proteins: uterine prealbumin, postalbumin, uteroglobin, uterine β-glycoprotein, and β-uterus-macroglobulin. On the other hand, albumin, transferrin, immunoglobulin IgG, and α-macroglobulin represent blood plasma proteins, which enter the uterine luminal fluid by selective transudation.

The secretory process of the specific endometrial proteins, their extrusion from uterine luminal epithelial cells, has received ample attention elsewhere (*Beier*, 1973, 1974a).

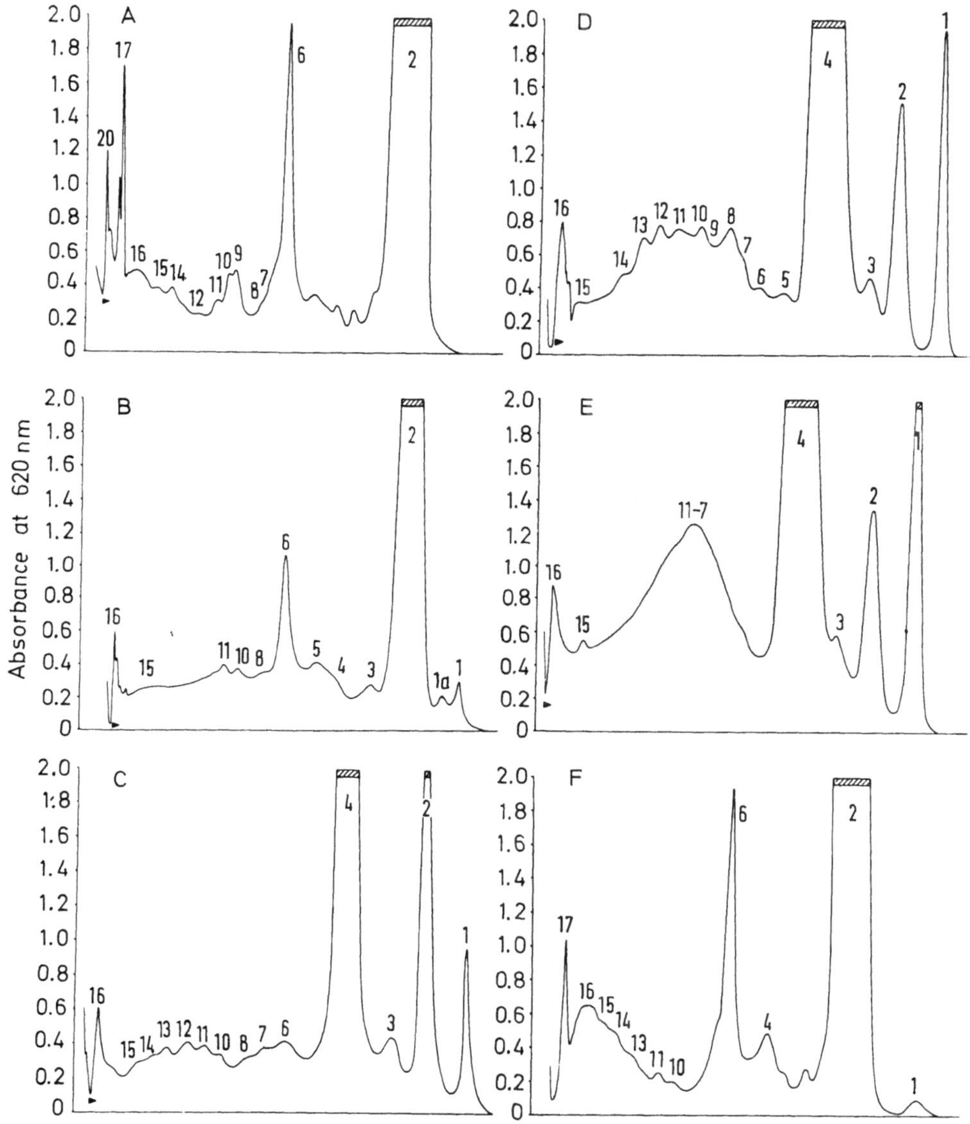

Fig. 4. PAGE protein patterns obtained after densitometry in transmitted light (Zeiss, PMQ III). Disc-electrophoresis was performed in 7.5% acrylamide, tris-glycine-buffer, pH 9.0. Amido black stained profiles are compared from rabbit blood serum (A), estrus uterine fluid in rabbit (B), uterine secretion of normal pregnancy at 4 days p.c. (C), 6 days p.c. (D), 7 days p.c. (E), and 9 days p.c. (F). Protein peaks are as follows: 1 Uterine prealbumin, 2 Albumin, 3 Uterine postalbumin, 4 Uteroglobin, 6 Transferrin, 7-11 Uterine β-glycoproteins, 17α-Macroglobulin (from *Beier*, 1973)

Uteroglobin is the prevailing fraction in rabbit endometrial secretion during the post-ovulation time until a brief period after implantation. This unique protein is a small

Table 1. Biochemical and endocrinologic parameters of secretion proteins in rabbit uterus

Protein	Components or properties	Molecular weight	Hormonal control in uterine secretion
Uterine prealbumin	Sialic acid-rich glycoprotein	ca. 50,000	Present in estrous and estrogen-treated animals. Increased secretion rate after progesterone influence
Albumin	No carbohydrate components. Osmotic regulation, transport protein.	68,000	Present in large amounts in estrous. Drastic decrease in consequence of progesterone secretion. Experimentally increased transudation after 17β-estradiol
Uterine postalbumin	Glucose-rich glycoprotein	ca. 20,000	Likely progesterone dependent
Uteroglobin	Sialic acid-free glycoprotein. Not yet clear embryotropic property. Binding of steroids, particularly progesterone	15,000	Not detectable in considerable amounts in estrous. After corpus luteum function is started, this protein increases in its relative and absolute amounts in secretion. Clear induction of uteroglobin after treatment of ovariectomized animals with progesterone or synthetic progestins
Transferrin	Glycoprotein. Binding and transport of iron.	80,000	Present in estrous, and increase in animals, which are treated with estrogens, according to increased capillary permeability
β-Glycoprotein	Galactose-rich glycoprotein. Proteolytic activity	70,000 –100,000	Progestagene dependency apparent after treatment of ovariectomized females with progesterone
Immunoglobulin IgG	Glycoprotein. Antibodies	150,000	Same as transferrin
α-Macroglobulin	Glycoprotein. Protease inhibitor	700,000 –800,000	Same as transferrin
β-Uterus-Macroglobulin	Highest molecular weight protein in uterine secretion	ca. 800,000 –1.000,000	Progesterone dependent protein, only detectable in pregnant and pseudopregnant uteri

macromolecule with a molecular weight of 14,000-15,000, composed most likely from two subunits of similar molecular size connected by a disulfide bridge (cf. Lit. *Beier*, 1974a; *Beato* and *Baier*, 1975). Uteroglobin has been detected in our laboratory (*Beier*, 1966, 1967) when comparative investigations on uterine fluid and blastocyst fluid protein patterns were carried out. Uteroglobin was described independently by *Krishnan* and *Daniel* (1967) and designated as "blastokinin" because of its yet questionable in vitro effects on rabbit blastocyst formation. As we have demonstrated (*Beier*, 1968a, b), this particular protein predominates the blastocyst fluid protein pattern by covering about 50% of the total protein amount during days 5 and 6 p.c. In the meantime, we have shown that uteroglobin, when found in blastocyst fluid, must have penetrated from the uterine luminal secretion into the blastocyst cavity. We have no evidence for uteroglobin synthesis in blastocyst tissues during in vitro growth of rabbit two-cell stages up to the ready-to-implant expanded blastocysts in a culture medium containing bovine serum albumin as the *only* protein source (*Beier* and *Maurer*, 1975), indicating that uteroglobin is not necessary for blastocyst expansion in vitro.

Uteroglobin, additionally, represents the major component in uterine fluid from pseudopregnant rabbits, and it can be isolated from the uterine secretion of progesterone-treated ovariectomized animals (*Beier*, 1968a, b; *Beier* et al., 1970). Physiological progestins can be replaced by synthetic steroids, exerting progestational activities, to evoke uteroglobin synthesis and secretion (*Beier* and *Beier-Hellwig*, 1973; *Beier* et al., in preparation). We have evidence, that the type of progestin is responsible for the quantitative hormone effect, leading to different quantitative secretion rates of uteroglobin in consequence of either different direct progestagen actions or different estrogenic partial activities of the synthetic steroids used. Besides the physiologically occurring steroids progesterone and 20α-dihydroprogesterone, we have found that chlormadinone acetate and d/l-norgestrel both are capable of inducing synthesis and extrusion of uteroglobin and the other pregnancy specific proteins in the rabbit uterus. The quantitative effect is mainly dependent on the pre-sensitization of the target tissue, i.e., the estrogen priming of the endometrium. Both synthetic progestins, the progesterone derivative 6-chloro-17-hydroxypregna-4,6-diene-3,20-dione acetate, as a typical C-19-methylated steroid, and the 19-nortestosterone derivative 13β-ethyl-17α-ethnyl-17β-hydroxy-4-gonen-3-one, as a typical 19-Nor-steroid, do stimulate uteroglobin secretion in castrated does which were primed with 17β-estradiol. The latter compound, d/l-norgestrel, also exerts a detectable effect without 17β-estradiol priming.

As demonstrated in Table 1, all secretory levels of the uterine luminal fluid proteins are controlled by ovarian hormones. These steroids regulate the steady state of the uterine secretion proteins. The pregnancy specific proteins are mainly under progestin control, whereas estrogens mainly are involved in the selective transudation process of proteins from the blood plasma into the uterine lumen. 17β-estradiol evidently influences the permeability of the endometrial surface epithelium and its underlying capillaries (cf. *Elger*, 1969; *Kühnel* et al., 1971; *Beier*, 1973).

IV. Experimentally Initiated Imbalances of the Endogenous Ovarian Hormone Levels

We have treated intact female rabbits, i.e., nonpregnant estrous does and animals after normal mating, with fertile bucks, in order to elucidate the effects of exogenously initiated imbalances of the endogenous ovarian hormone levels on the uterine secretion proteins.

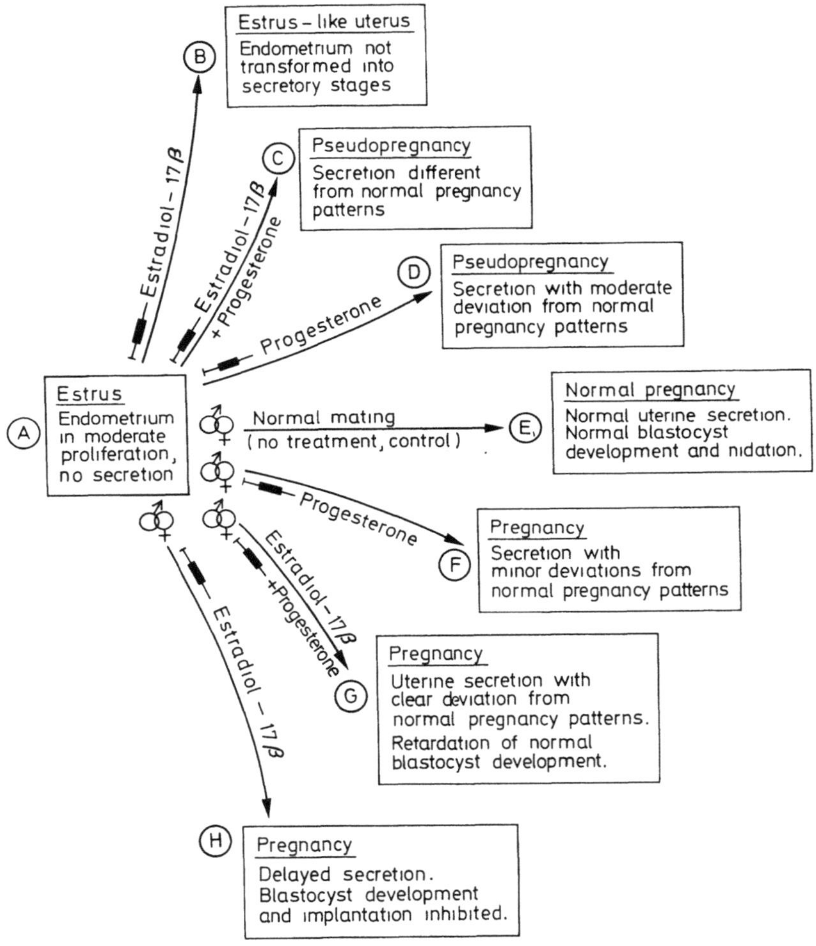

Fig. 5. Design of experiments with hormonal imbalances, induced by exogenous ovarian steroids in pseudopregnant and normal pregnant rabbits. Control groups are estrus animals (A) and normal pregnant does after normal mating without steroid treatment (E). All experiments started with normal estrus animals and each analysis performed on day 6 p.c. Groups B, C, D only received hormone injections without preceding copulations, thus leading to different pseudopregnant stages, one of which (B) was not comparable with real pseudopregnancy. Hormone treatments of experiments F, G, H performed after natural mating. Dosage of exogenously applied hormones: 17β-estradiol, intramuscularly (i.m.) injections of 0.1 mg at 6 h p.c. and 0.15 mg 30 h p.c.; progesterone, i.m. injections of 5.0 mg at days 1 to 5 p.c. each. (Compiled from *Beier* et al., 1970; *Beier*, 1974a)

One of the primarily envisaged problems was the comparatively complex question on contraceptive steroid effects at the uterine site, studied in order to establish an animal model of rabbit reproduction. With regard to the interesting feature of normal ovulations, which occur during treatment with the so-called ovulation-inhibitors (the "pill") in normal cycling women, considerable discussions emerged from the problematical contraceptive steroid effects after ovulation and fertilization. As has been earlier discussed by *Chang* (1966, 1967, 1969), these orally applied contraceptive steroids are so extraordinarily effective because they most likely disturb many reproductive processes, besides the inhibition of ovulation. The inhibition of nidation is generally the final event if several components are working in the chain reaction of pregnancy prevention by the pill. From our own investigations, evidence can be presented that the uterine secretion proteins are likely the most important factors which are disturbed by the exogenous hormonal interference in the complex system that permits normal implantation of the blastocyst in a properly conditioned uterus.

Figure 5 outlines an experimental schedule by which we have tested the response of the endometrial protein synthesizing system to different exogenous elevations either of the estrogen levels, the progesterone levels, or both of them. Such induced endocrine imbalances reveal variations in the secretory protein patterns, and reflect the inability of the maternal genital tract to produce these uterine proteins at the appropriate time, when the blastocyst needs them. The hormonal treatment of normal estrous animals without mating, as described in experiments C and D, yields abnormal pseudopregnancy patterns. In the course of the experiments where 17β-estradiol was applied to normal pregnant animals (Exps. G and H), blastocyst development has been retarded or even totally stopped, consequently no implantation could occur. By day 6 p.c., protein patterns of the disturbed pregnancies show remarkable variations compared to the normal. In summary, 17β-estradiol exerts the most remarkable effect among the ovarian hormones tested on the development and dynamic of the protein pattern in the uterine secretion (*Beier* et al., 1970, 1971). This steroid, given between the day of mating and day 3 p.c. in a single injection of 50 μg or more, up to 250 μg, stimulates high concentrations of serum-identical proteins, and, on the other hand, initiates the suppression of uterine-specific protein secretion (*Beier* et al., 1970). All estrogen-dominated experimental situations (Exps. A, B, H) reveal high proportions of albumin, the prevailing serum protein, but there appear low levels when the uterine secretion is controlled by progesterone (Exps. C, D, E, F, G), as shown in Figure 6. The reverse situation with uteroglobin concentrations and proportions, respectively, is indicated by Figure 7.

Summarizing these effects, we must conclude that the endometrial secretion is disturbed, to an extent dependent upon the dosage and frequency of the exogenously applied ovarian steroids. Parallel to the distorted secretion picture, various forms of retarded blastocyst development can be recognized. In order to shed more light on the questionable causal relationship between the endometrial failure in protein secretion and the blastocyst retardation, we have paid particular attention to the analysis of the 17β-estradiol-treatment shortly after normal mating. As described earlier, the longitudinal study of this treatment's effects from day 4 to day 18 p.c. on the site of uterine secretion yields a considerably delayed morphological endometrial transformation, as well as a particularly delayed protein extrusion of the secretory uterine epithelial cells (*Beier* et al., 1971; *Beier* and *Kühnel*,

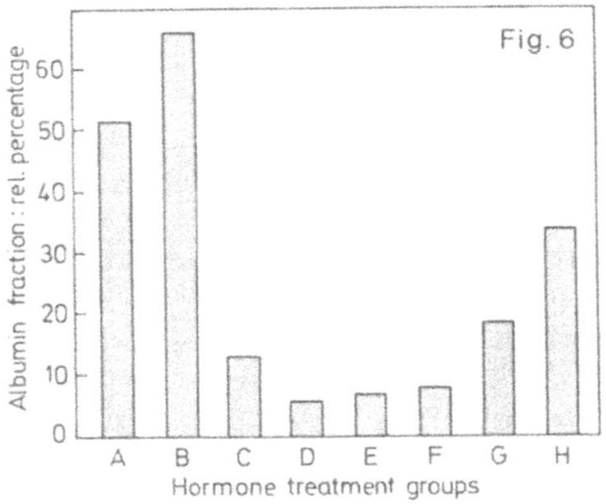

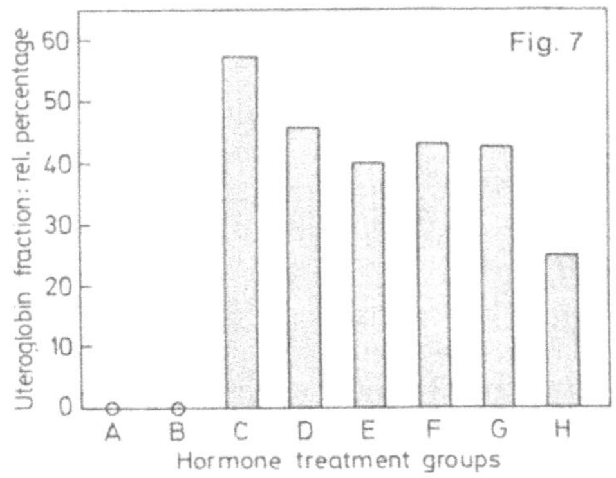

Fig. 6. Relative percentage of albumin fraction in rabbit uterine secretion of hormone-treated animals from exps. A to H, as outlined in Figure 5. Percentage values are mean values obtained from pooled specimens after densitometry of discelectrophoretical columns, as method shown in Figure 4. (Modified diagram from *Beier*, 1974a)

Fig. 7. Relative percentage of uteroglobin fraction in rabbit uterine secretion of hormone-treated animals from Exps. A to H, as outlined in Figure 5. Percentage values are mean values obtained from pooled specimens after densitometry of discelectrophoretical columns. Exps. A and B reveal amounts of uteroglobin not detectable by method used. (Modified diagram from *Beier*, 1974a)

1972; *Beier*, 1973, 1974a). Consequently, I have designated this phenomenon "delayed secretion." Results of the biochemical investigations on the protein pattern dynamics during delayed secretion are shown in Figures 8, 9, and 10.

Interestingly, we could demonstrate these hormone effects clearly in the uterus, but not in the oviduct. This finding reflects a remarkable difference between two target tissues, which usually are thought to be controlled by the same estrogen/progestin ratios and, consequently, were expected to respond similarly. The same differential hormone effect could be substantiated further by morphologic findings (*Kühnel* and *Beier*, 1975).

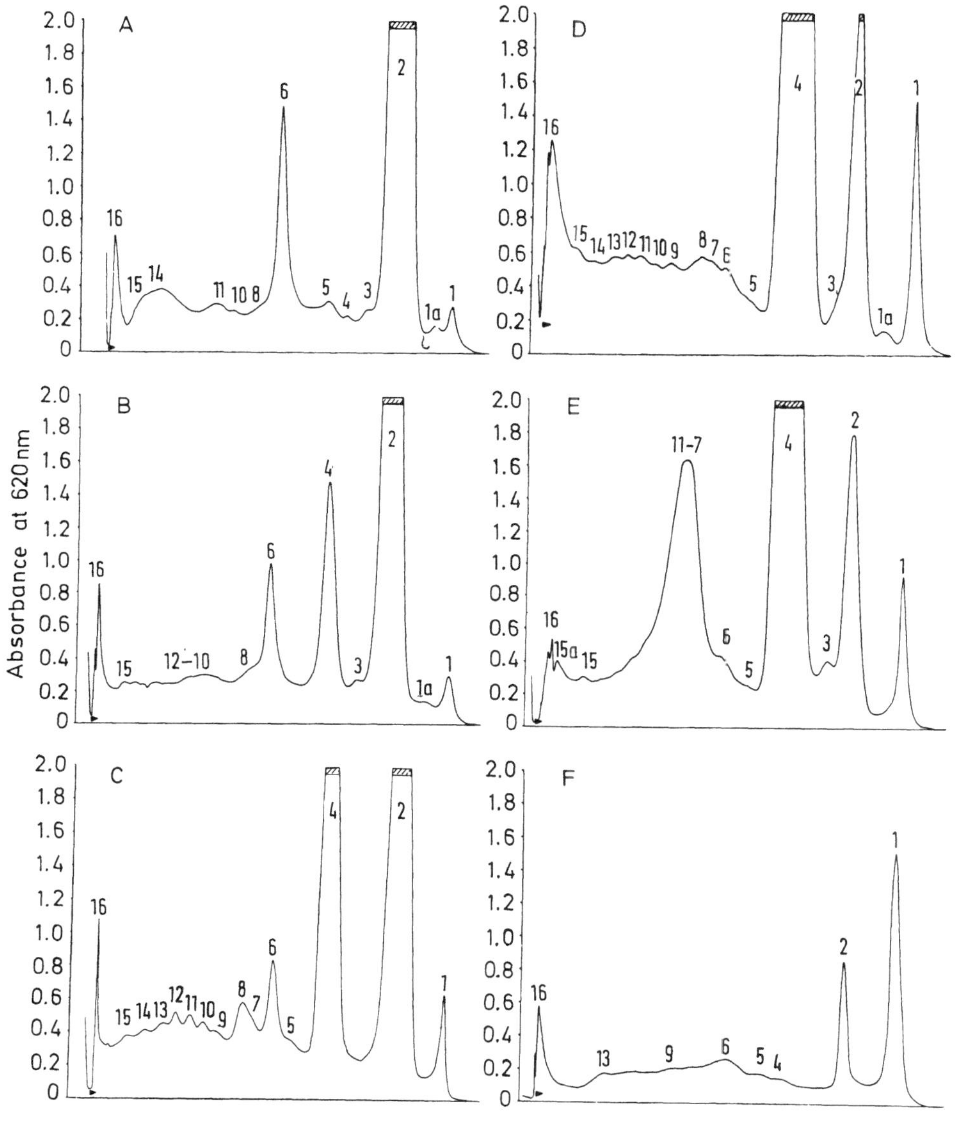

Fig. 8. PAGE protein patterns (obtained as described in Figure 4) from postcoitally 17β-estradiol treated rabbits, showing delayed secretion. (A) Control treatment of estrus animals yields no significant change of estrus uterine fluid pattern. Experimental patterns obtained at day 4 p.c. (B), day 6 p.c. (C), day 8 p.c. (D), day 12 p.c. (E), and day 18 p.c. (F) show delayed protein pattern development compared to normal pattern dynamic. (From *Beier*, 1973)

Fig. 9

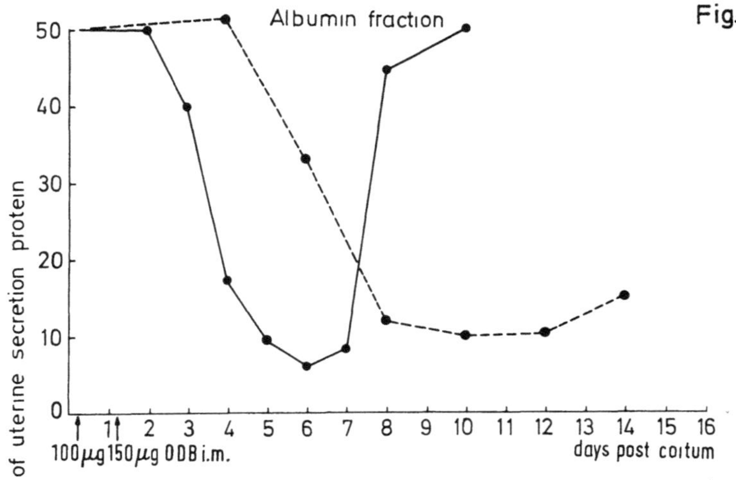

Fig.10

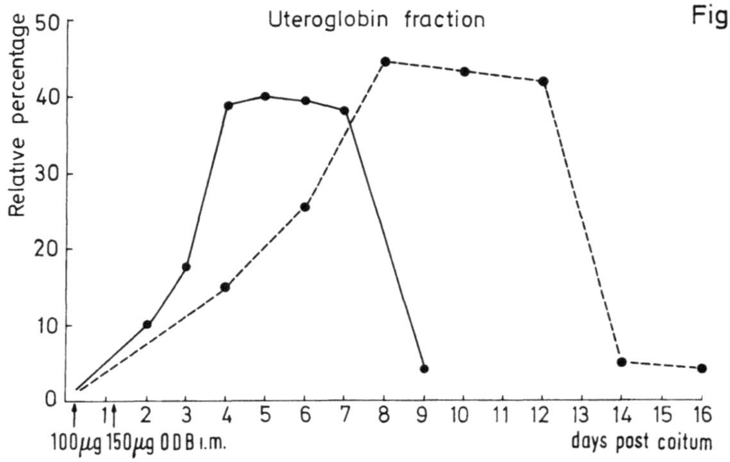

Fig. 9. Relative percentage of albumin fraction in rabbit uterine fluid throughout delayed secretion of endometrium. Mean values obtained from pooled specimens after densitometry of disc-electrophoretical columns. Proportions from normal pregnancies (black line) are compared with delayed secretions (dotted line). Estradiol injections indicated by arrows. ODB = 17β-estradiol-benzoate. (From *Beier*, 1973).

Fig. 10. Relative percentage of uteroglobin fraction in rabbit uterine fluid throughout delayed secretion of endometrium. Mean values obtained from pooled specimens after densitometry of disc-electrophoretical columns. Proportions from normal pregnancies (black line) compared with delayed secretions (dotted line). Estradiol injections indicated by arrows. ODB = 17β-estradiol-benzoate. (From *Beier*, 1973)

V. Biological Significance of Uterine Secretion Proteins and Their Hormonal Control

Today there is no clear information available, as to what concerns the biological significance of the proteins observed in the compartments of the endometrium and the unimplanted blastocyst. Recent findings indicate that uteroglobin is a carrier protein for steroids,preferably for progesterone (*Beato* and *Baier*, 1975). If we suppose this progesterone binding by uteroglobin as its real function, and additionally, consider the well-known binding of estradiol to albumin, we find the establishment of a steroid-carrier protein system, which supplies the developing blastocyst with hormones from the maternal endocrine system (*Beier*, 1974b). Several proteins of uterine secretion have been described and analyzed as enzymes, the significance of which is still obscure in the developmental process of the blastocyst.

Reviewing the literature (*Maurer* et al., 1970; *van Blerkom* et al., 1973; *Jenkinson* and *Wilson*, 1973; *van Blerkom* and *Manes*, 1974; *Salomon* and *Sherman*, 1975) and taking into account recent results of our own laboratory (*Beier* and *Maurer*, 1975), for comparative evaluation of in vitro versus in vivo development it becomes quite evident that the blastocyst's differentiation occurs under entirely different conditions. The possibility of in vitro culturing rabbit blastocysts from the early cleaving stages to the expanded blastocyst stages without any need for uterine-specific proteins in the culture medium, leads us to conclude that steroid hormones and their binding proteins regulate blastocyst development and implantation in utero, but not in vitro. The complex system of the maternal environment in utero only permits implantation within a very limited period, whereas at extrauterine sites apparently implantations may occur at any time and condition, e.g., in the anterior chamber of the eye within immature males (*Runner*, 1947). Blastocysts in utero never develop in the absence of estrogens and progestins. Asynchronous egg transfer in the normal and in the delayed secretion uteri of rabbits has clearly shown the essential feature of embryo-maternal synchronization requirements in utero (*Beier* et al., 1972; *Beier*, 1973, 1974a). Finally, our conclusion may be, that the significance of ovarian hormonal control during preimplantational mammalian development comprises more of a maternal than an embryonic target.

Acknowledgments: The support from the Deutsche Forschungsgemeinschaft is gratefully acknowledged. Unpublished results of this paper from the author's laboratory have been supported by DFG-grants Be 524/2 and Be 542/4. In addition, the author acknowledges with many thanks the substantial support by the Schering AG, Berlin, particularly from Prof. Dr. *Friedmund Neumann* and Dr. *Walter Elger*, Dept. of Endocrinpharmacology.

References

Baldwin, D.M., Stabenfeldt, G.H.: Plasma levels of progesterone, cortisol, and corticosterone in the pregnant rabbit. Biol. Reprod. 10, 495-501 (1974).

Beato, M., Baier, R.: Binding of progesterone to the proteins of the uterine luminal fluid: Identification of uteroglobin as the binding protein. Biochim. biophys. Acta (Amst.) 392, 346-356 (1975).

Beier, H.M.: Das Proteinmilieu in Serum, Uterus und Blastocysten des Kaninchens vor der Nidation. Coll. Biochemie d. Morphogenese, Dtsch. Forsch.-Gem., Konstanz 1966.

Beier, H.M.: Veränderungen am Proteinmuster des Uterus bei dessen Ernährungsfunktion für die Blastocyste des Kaninchens. Verh. dtsch. zool. Ges., Heidelberg 31, 139-148 (1967).

Beier, H.M.: Biochemisch-entwicklungsphysiologische Untersuchungen am Proteinmilieu für die Blastocystenentwicklung des Kaninchens (Oryctolagus cuniculus). Zool. Jb. Anat. 85, 72-190 (1968a).

Beier, H.M.: Uteroglobin: A hormone-sensitive endometrial protein involved in blastocyst development. Biochim. biophys. Acta (Amst.) 160, 289-291 (1968b).

Beier, H.M.: Protein patterns of endometrial secretion in the rabbit. In: Ovo-Implantation, Human-Gonadotropins and Prolactin (eds. *P.O. Hubinont* et al.), pp. 157-163. Basel–München–New York: S. Karger 1970.

Beier, H.M.: Die hormonelle Steuerung der Uterussekretion und frühen Embryonalentwicklung des Kaninchens. Habil.-Schrift, Med. Fakultät Univ. Kiel, pp. 1-216, 1973.

Beier, H.M.: Oviducal and uterine fluids. J. Reprod. Fertil. 37, 221-237 (1974a).

Beier, H.M.: Ovarian steroids in embryonic development before nidation. Advanc. Biosci. 13, 199-219 (1974b).

Beier, H.M., Beier-Hellwig, K.: Specific secretory protein of the female genital tract. Acta endocrin. (Kbh.) Suppl. 180, 404-425 (1973).

Beier, H.M., Elger, W., Kühnel, W., Neumann, F.: Beeinflussung der Uterussekretion des Kaninchens durch synthetische Gestagene (in preparation).

Beier, H.M., Kühnel, W.: Die verzögerte Uterussekretion nach Oestrogeninjektionen beim graviden Kaninchen. Verh. anat. Ges. Köln 1972, Anat. Anz. 134 (Suppl.), 567-575 (1973).

Beier, H.M., Kühnel, W., Petry, G.: Uterine secretion proteins as extrinsic factors in preimplantation development. Advanc. Biosci. 6, 165-189 (1971).

Beier, H.M., Maurer, R.R.: Uteroglobin and other proteins in rabbit blastocyst fluid after development in vivo and in vitro. Cell Tissue Res. 159, 1-10 (1975).

Beier, H.M., Mootz, U., Kühnel, W.: Asynchrone Eitransplantationen während der verzögerten Uterussekretion beim Kaninchen. Proc. VII. Intern. Congr. Animal Reprod. and Artif. Insemin., München 1972, Bd. III, 1891-1896 (1972).

Beier, H.M., Petry, G., Kühnel, W.: Endometrial secretion and early mammalian development. In: 21. Coll. Ges. Biol. Chem. Mosbach 1970 (eds. *H. Gibian, E.J. Plotz*), pp. 264-285. Berlin–Heidelberg–New York: Springer 1970.

Blerkom, J. van, Manes, C., Daniel, J.C., jr.: Development of preimplantation rabbit embryos in vivo and in vitro. I. An ultrastructural comparison. Develop. Biol. 35, 262-282 (1973).

Blerkom, J. van, Manes, C.: Development of preimplantation rabbit embryos in vivo and in vitro. II. A comparison of qualitative aspects of protein synthesis. Develop. Biol. 40, 40-51 (1974).

Bullock, D.W., Willen, G.F.: Regulation of a specific uterine protein by estrogen and progesterone in ovariectomized rabbits. Proc. Soc. exp. Biol. Med. 146, 294-289 (1974).

Challis, J.R.G., Davies, I.J., Ryan, K.J.: The concentrations of progesterone, estrone and estradiol-17β in the plasma of pregnant rabbits. Endocrinology 93, 971-976 (1973).

Chang, M.C.: Effects of oral administration of medroxprogesterone acetate and ethinyl estradiol on the transportation and development of rabbit eggs. Endocrinology 79, 939-948 (1966).

Chang, M.C.: Effects of progesterone and related compounds on fertilization, transportation and development of rabbit eggs. Endocrinology 81, 1251-1260 (1967).

Chang, M.C.: Fertilization, transportation and degeneration of eggs in pseudopregnant or progesterone-treated rabbits. Endocrinology 84, 356-361 (1969).

Davies, I.J., Ryan, K.J.: Comparative endocrinology of gestation. Vitam. Horm. 30, 223-279 (1972).

Eaton, L.W., Hilliard, J.: Estradiol-17β, progesterone and 20α-hydroxypregn-4-en-3-one in rabbit ovarian venous plasma. I. Steroid secretion from paired ovaries with and without corpora lutea, effect of LH. Endocrinology 89, 105-111 (1971).

Elger, W.: Einfluß von Gestagenen auf Befruchtung, Eiernährung, Eitransport und Schwangerschaftsverlauf. In: Hdb. exp. Pharmakologie, N.S. (eds. *O. Eichler* et al.), Bd. XXII/2 (ed. *K. Junkmann*), pp. 132-284. Berlin–Heidelberg–New York: Springer 1969.

Fuchs, A.R., Beling, C., Park, K.: Corpus luteum recognition of blastocysts prior to implantation. 56th Meeting, The Endocrine Soc. June 1974, Endocrinology 95, A-167 (1974).

Hafez, E.S.E.: Growth and survival of blastocysts in the domestic rabbit. II. Quantitative effects of exogenous progesterone following ovariectomy. J. Reprod. Fertil. 7, 241-249 (1964).

Hafez, E.S.E., Pincus, G.: Hormonal requirements of implantation in the rabbit. Proc. Soc. exp. Biol. Med. 91, 531-534 (1956).

Hasan, S.H., Neumann, F., Friedreich, E., Elger, W.: Levels of progesterone during pregnancy in rabbits as measured by radioimmunoassay. Acta endocr. (Kbh.) Suppl. 155, 107 (1971).

Hilliard, J.: Corpus luteum function in guinea pigs, hamster, rats, mice, and rabbits. Biol. Reprod. 8, 203-221 (1973).

Hilliard, J., Eaton, L.W.: Estradiol-17β, progesterone and 20α-hydroxypregn-4-en-3-one in rabbit ovarian venous plasma. II. From mating through implantation. Endocrinology 89, 522-527 (1971).

Hilliard, J., Pang, C.-N., Scaramuzzi, R.J., Penardi, R., Sawyer, C.H.: Secretion rates of estradiol, testosterone, and progesterone from right and left ovaries cannulated concurrently or successively. Biol. Reprod. 10, 364-369 (1974).

Hilliard, J., Spies, H.G., Sawyer, C.H.: Cholesterol storage and progestin secretion during pregnancy and pseudopregnancy in the rabbit. Endocrinology 82, 157-165 (1968).

Jenkinson, E.J., Wilson, I.B.: In vitro studies on the control of trophoblast outgrowth in the mouse. J. Embryol. exp. Morph. 30, 21-30 (1973).

Kirchner, C.: Untersuchungen an uterusspezifischen Glycoproteinen während der frühen Gravidität des Kaninchens (Oryctolagus cuniculus). Wilhelm Roux' Arch. Entwickl.-Mech. Org. 164, 97-133 (1969).

Kirchner, C., Hirschhäuser, C., Kionke, M.: Protease activity in rabbit uterine secretion 24 hours before implantation. J. Reprod. Fertil. 27, 259-260 (1971).

Krishnan, R.S., Daniel, J.C.: "Blastokinin": Inducer and regulator of blastocyst development in the rabbit uterus. Science 158, 490-492 (1967).

Kühnel, W., Beier, H.M., Petry, G.: Untersuchungen zur hormonellen Regulation der Praeimplantationsphase der Gravidität. II. Histologische, topochemische und biochemische Analysen am hormonbehandelten, pseudograviden Kaninchenuterus. Cytobiologie 4, 9-40 (1971).

Lutwak-Mann, C., Hay, M.F., Adams, C.E.: The effect of ovariectomy on rabbit blastocysts. J. Endocr. 24, 185-197 (1962).

Maurer, R.R., Onuma, H., Foote, R.H.: Viability of cultured and transferred rabbit embryos. J. Reprod. Fertil. 21, 417-422 (1970).

Mikhail, G., Noall, M.W., Allen, W.M.: Progesterone levels in the rabbit ovarian vein blood throughout pregnancy. Endocrinology 69, 504-509 (1961).

Okano, K., Matsumoto, K., Kotoh, K., Endo, H.: Progestins in the ovarian vein blood of non-pregnant and pregnant rabbits before and after gonadotropic stimulation. Endocr. jap. 13, 438-447 (1966).

Runner, M.N.: Development of mouse eggs in the anterior chamber of the eye. Anat. Rec. 98, 1-17 (1947).

Salomon, D.S., Sherman, M.I.: Implantation and invasiveness of mouse blastocysts on uterine monolayers. Exp. Cell Res. 90, 261-268 (1975).

Shaikh, A.A., Harper, M.J.K.: Ovarian steroid secretion in estrous, mated and HCG-treated rabbits, determined by concurrent cannulation of both ovarian veins. Biol. Reprod. 7, 387-397 (1972).

IV. Teratology

Organ Culture in Teratology

L. SAXÉN, M. KARKINEN-JÄÄSKELÄINEN, and I. SAXÉN

I. Introduction

Our present, still fragmentary knowledge of teratogens and teratogenesis is based on studies following two major methodologic lines, non-experimental, epidemiologic studies, and experimental, in vivo investigations. Both approaches are hampered by many limitations and shortcomings, the most obvious in the epidemiologic studies being the complicated study situation and the great number of confounding factors (*Saxén*, 1974), whereas the greatest difficulty in animal experiments is the inapplicability of the results to other species, including human embryos (*Fraser*, 1964). Hence, we need to explore alternative and complementary methods for testing various substances for teratogenicity and for studying the mechanisms of abnormal development. Tissue culture has now been used for decades in many fields of biomedical research and has proved a most valuable tool in the hands of skillful, critical scientists. In teratology, the method may also eliminate some of the difficulties and shortcomings of epidemiological and in vivo studies, but so far it has found only a limited number of applications in this field. Cell culture methods have recently become indispensable in diagnostic teratology but their use in basic studies is difficult. Here organ culture, which allows good differentiation, morphogenesis, and often functional maturation of embryonic tissues in vitro, might prove more useful.

In what follows, some introductory comments on the advantages and disadvantages of the organ culture technique will be made and the basic methods briefly presented. Thereafter three examples of our own work will be briefly described. These will illustrate the use of organ culture methods in studies representing three categories of basic problems in teratology: the mode of action of exogenous teratogens, the problem of the "sensitive periods" of development, and the interaction between genetic and environmental factors in teratogenesis.

1. Advantages and Disadvantages of the Organ Culture Method

Studies on developing organ rudiments or pieces of embryonic tissues in vitro offer several advantages for teratologic investigations as compared to in vivo experiments:

1. Confounding maternal factors are eliminated.
2. Placental barriers, which vary between species, are bypassed.
3. The experimental conditions are stable and reproducible.
4. The target cells and tissues can be easily and precisely exposed to the factors studied.
5. The target tissue can be continuously observed during the experiment.
6. The stage of development of the target organ can be precisely determined at the time of exposure, and when paired organs are used, the control culture can be set up from the same embryo.
7. Human embryonic tissues can be used for direct experimental studies whereas in animal experiments interspecific differences in the responses to particular teratogens reduce the clinical value of the results.

On the other hand, a teratologist choosing the organ culture method for his studies should be aware of certain grave disadvantages of this technique and the artificial conditions of such experiments before extrapolating his results to the in vivo situation:

1. The organismal control systems and various protective mechanisms are eliminated.
2. Excision and cultivation of embryonic organs inevitably interfere with their development, and artifacts due to these causes have to be distinguished from the true effects of the teratogen under examination.
3. The technique itself has several limitations such as the size of the piece of tissue, the length of the culture period, the stage of development that can be reached, etc.

2. Comments on the Basic Methods

The basic technique for culturing embryonic organs or pieces of tissue was introduced half a century ago (*Strangeways* and *Fell*, 1926) but only developed into a routine method in the fifties, with the advent of antibiotics. Many modifications of the basic technique have since been devised (see *Fell*, 1951, 1961; *Trowell*, 1961; *Wessells*, 1967, and others); two of them are illustrated in Figure 1. The tissues are usually grown either on a solid nutrient

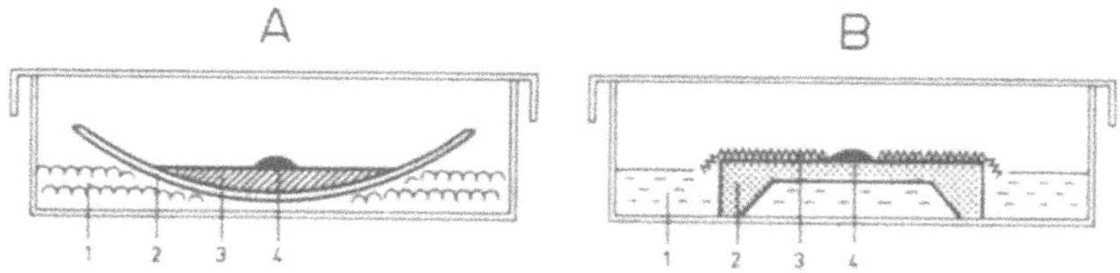

Fig. 1. Scheme of two common methods of organ culture: A. Watchglass technique of *Fell* (1951: 1: wet cotton support, 2: watchglass, 3: solid nutrient, 4: tissue. B: Screen method of *Trowell* (1961): 1: liquid nutrient medium, 2: stainless stell screen, 3: filter paper, 4: tissue

or at the interphase between a gaseous atmosphere and a liquid culture medium. Various commercial Petri dishes with supportive structures can be used, but specially designed "organ culture dishes" are also available (*Wessells*, 1967). The composition of the nutrient medium has developed into a science of its own with both unknown and commercial components. The final choice of a medium for a particular study and tissue is ultimately based on the experiences of the research worker himself, and there is no need for any detailed discussion of that problem here. Instead, we shall give a brief account of the methods used in our own studies, some of which are described below.

The basic technique follows that introduced by *Trowell* (1961), and is illustrated in Figure 1B: The tissue grows at the medium/gas interface supported on a metal screen by a strip of filter paper. The humidified atmosphere consists of 5% CO_2 in air. In most of our studies on mouse tissues, a commercial "minimum essential medium" (GIBCO, New York, Cat. No. F-17) has been used, supplemented with 10% inactivated fetal calf serum (*Saxén* et al., 1968). In two of the studies to be described here, a chemically defined medium without protein supplementation was used. This medium, BGJb (*Biggers* et al., 1961), allows good growth and differentiation of bone rudiments (provided that ascorbate is added daily), and supports differentiation and fusion of mouse palatal shelves (see below). In the third series of experiments, in which chick embryonic lens was cultivated, a variety of media were tested, and an exceptionally high protein content was found to be optimal (20% serum plus 20% chick embryo extract) (*Karkinen-Jääskeläinen*, 1973). For human tissues, various commercial media with added serum were found suitable (*Lash* and *Saxén*, 1972), but in some cases human amniotic fluid without any addition gave excellent results (unpublished).

II. Three Examples of Problems Approached by Organ Culture

1. The Mode of Action of Teratogenic Drugs

a) Introduction

Tetracycline was chosen for the model substance of chemical teratogenesis for several reasons: This widely used antibiotic is easily transmitted from the maternal circulation to the fetal organism (*Charles*, 1954; *Simpson* et al., 1967), where it is taken up by the skeletal tissues, especially at sites of active mineralization (*Milch* et al., 1958; *Bevelander* et al., 1961) (Fig. 2). It also interferes with bone and tooth development in immature tissues of animals, including man (*Bevelander* et al., 1960; *Wallman* and *Hilton*, 1962; *Cohlan* et al., 1963; and others). The mechanism by which the drug inhibits bone development was not known, and we therefore started a series of experiments in vitro to study the matter.

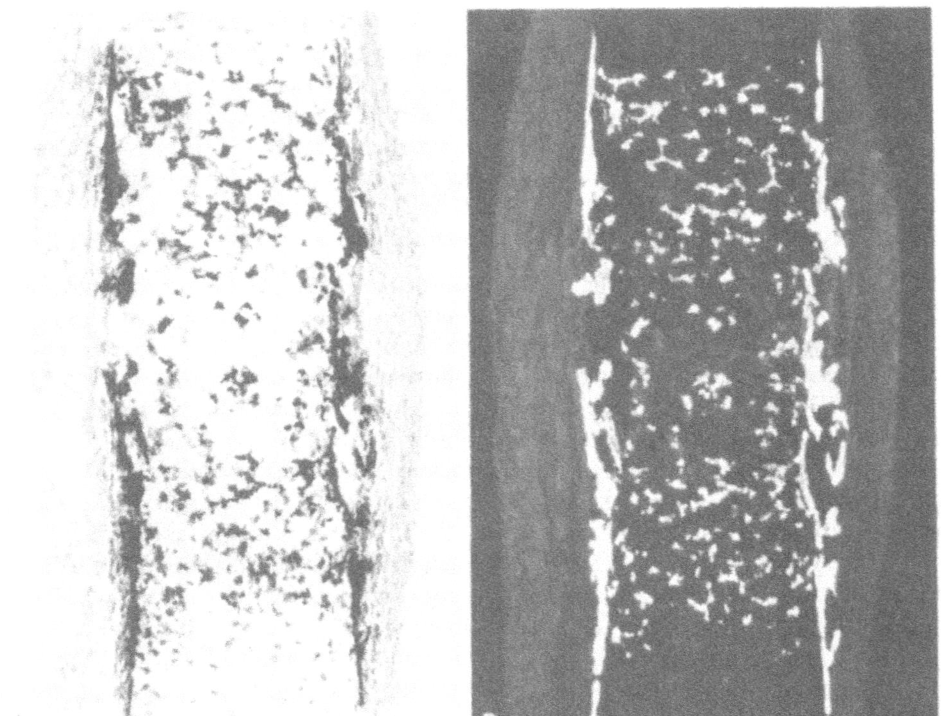

Fig. 2. Section of 16-day mouse embryonic radius cultivated in presence of tetracycline hydrochloride. A. Same section after von Kossa's stain for calcium. (*Saxén*, 1966). B. Section photographed in UV-light, demonstrating distribution of tetracycline fluorochrome

Long bones of 16 day mouse embryos were grown in a Trowell-type culture in the chemically defined medium already mentioned, and their development was followed by morphological, biochemical, and radiochemical methods. In the presence of tetracycline in concentrations comparable to those obtained in clinical work, inhibition of development could be demonstrated by following the elongation of the mineralized zone of the bones and

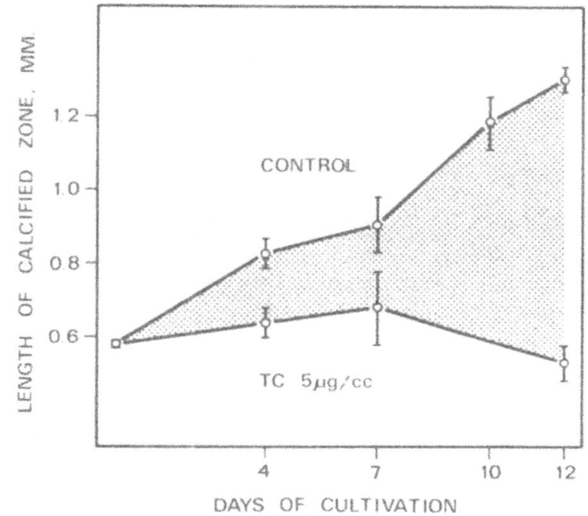

Fig. 3. Effect of tetracycline on elongation of mineralized zone of mouse embryonic ulnae in vitro. (After *Saxén*, 1966)

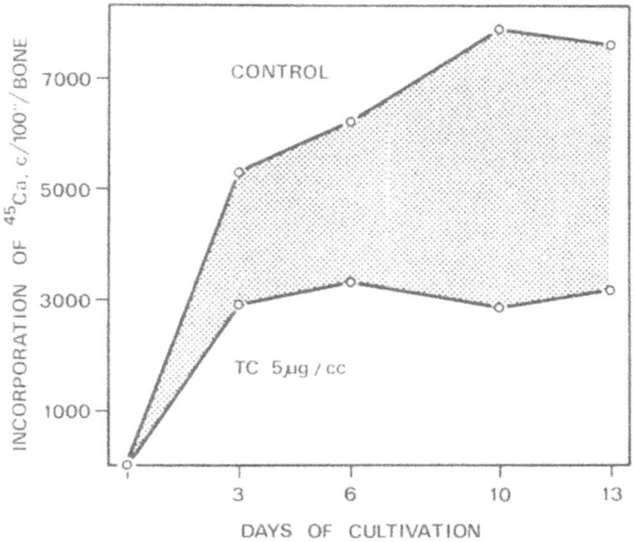

Fig. 4. Effect of tetracycline on cumulative uptake of radiocalcium by mouse embryonic ulnae in vitro. (After *Kaitila* et al., 1970)

measuring the cumulative uptake of radiocalcium (*Saxén*, 1965, 1966; *Kaitila* et al., 1970). When tetracycline hydrochloride was present in a concentration of 5 μg/ml, bone formation came to a standstill after a period of 3-5 days in culture (Figs. 3 and 4).

The target site of the teratogen still remained unknown, and might have been any step in the chain of events starting from the differentiation and proliferation of the chondroblasts, continuing with the synthesis of the cartilaginous matrix (mucopolysaccharides and collagen), and ending with the formation and growth of the bone mineral crystals. For identification of the target site(s) and understanding of the mode of action of tetracycline, a definite effect on one of these steps had to be detected, and inhibition of the others excluded.

b) Results

Proliferation of the chondroblasts was analyzed by following the incorporation of tritiated thymidine into the bones in the presence of various concentrations of tetracycline. High tetracycline concentrations (100 μg/ml did inhibit this incorporation, whereas lower levels (1 to 10 μg/ml) had no detectable effect on the proliferation thus estimated, and yet bone formation was shown to be inhibited at these concentrations (*Kaitila* et al., 1970). Next, an effect on the synthesis of mucopolysaccharides was excluded in a similar way by showing that incorporation of labeled SO_4 proceeded at the control level for up to 12 days in the presence of tetracycline at concentrations effectively inhibiting bone formation (*Kaitila* et al., 1970). The rate of collagen synthesis was evaluated by following the uptake of radioactive proline and its hydroxylation in cultured bones. Inhibition of collagen biosynthesis by tetracycline was demonstrated with high tetracycline concentrations or after prolonged culture, but not with concentrations or at a stage of cultivation where bone formation was already impaired (Table 1). Further experiments suggested that the action of tetracycline in high concentrations was due to the binding of the ferrous iron required for collagen biosynthesis (*Halme* et al., 1969).

Table 1. Effect of various concentrations of tetracycline on collagen biosynthesis in embryonic bones cultured in vitro in the presence of 14-C-proline. (*Saxén* and *Kaitila*, 1972, after *Halme* et al., 1969)

Tetracycline μg/ml	14-C-Hydroxyproline dpm percent of control		Hydroxylation percentage	
	day 3	day 6	day 3	day 6
None	100	100	22.3	20.0
10.0	95	67	20.8	20.1
30.0	91	76	20.8	18.0
100.0	–	50	–	13.7

Further evidence for the conclusion that tetracycline in low concentrations does not exert its effect by inhibiting protein synthesis was obtained in experiments where the drug was tested on bone rudiments whose protein synthesis was completely blocked by cycloheximide (*Kaitila*, 1971a). In short-term cultures, uptake of radiocalcium proceeded at the control level after protein synthesis was inhibited, and the effect of tetracycline on these bones did not differ from that of the control cultures with unaffected protein synthesis (Table 2).

Table 2. Effect of tetracycline on incorporation of radiocalcium into bones with blocked protein synthesis and in those with normal synthesis (*Saxén* and *Kaitila*, 1972; after *Kaitila*, 1971a).

Treatment	45-Calcium dpm percent of control
Tetracycline (10 μg/ml)	64
Tetracycline (10 μg/ml) and cycloheximide (2 μg/ml)	66

The final study was concerned with the effect of tetracycline on the last step in bone development, i.e., the formation and growth of the mineral crystals (*Kaitila*, 1971b). Use was made of the in vitro technique of *Katz* (1969), where bone mineral formation is induced in a buffer solution by reconstituted collagen. Mineral formation was evaluated either by measuring the amount of calcium and phosphate in the collagen or by following the elimination of these from the buffer solution. The results, some of which are illustrated in Figure 5, show that tetracycline hydrochloride at a concentration of 5 μM (2.4 μg/ml) or more delayed the elimination of Ca^{++} from the medium. In further studies employing the kinetic model of *Katz* (op. cit.) a distinction could be drawn between an effect on the formation of bone mineral crystal and an inhibition of its subsequent growth. It was concluded that in the presence of 20 μM tetracycline, the first step was inhibited by a factor of 35% and the growth phase by 70% (*Kaitila*, op. cit.).

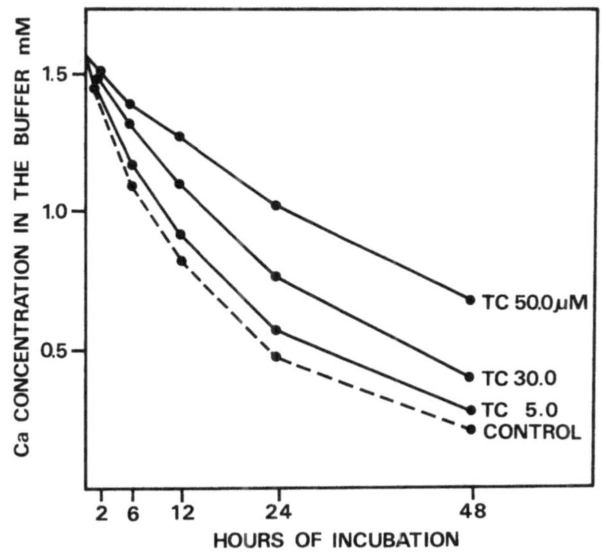

Fig. 5. Effect of various concentrations of tetracycline on elimination of calcium from buffer solution containing reconstituted collagen as inductor of bone mineralization. (After *Kaitila*, 1971b)

c) Conclusions

These experiments led us to conclude that the primary site of action of tetracycline in bone development is the formation and growth of the mineral crystal. Proliferation of chondroblasts and synthesis of the matrix components are not affected by tetracycline in low concentration. It has been suggested that the action of tetracycline molecules might be due to competition with the divalent cations of the bone mineral, but there is no direct evidence for this (*Kaitila*, 1971b; *Saxén* and *Kaitila*, 1972).

Finally, we should consider whether our results obtained under in vitro conditions and conclusions can be extrapolated to the in vivo situation, and, perhaps, to the "clinical" object, the human fetus. We would like to stress that these particular in vitro studies demonstrated the effect of a teratogenic drug on a basic and general developmental process, the formation and growth of bone mineral crystals. Assuming that this physiochemical process is similar in different vertebrate species in vivo, we could generalize the conclusions. Consequently, we would have in tetracycline a relatively well-established chemical teratogen with a known target site and mode of action.

2. Mechanism of Sensitive Periods to Exogenous Agents

a) Introduction

Embryonic development is characterized by sensitive periods during early organogenesis. At these times, the immature organism is poorly buffered to exogenous teratogens, whether physical or chemical, and congenital malformations often ensue. Susceptibility to viral damage frequently undergoes gradual restriction from destruction of all the cells of a very young embryo to compartmentalization of the damage to certain organs or cell populations in stages of sensitivity, each with its own narrow temporal and spatial limits. Thereafter it reaches a level where reactions are comparable to those of the adult organism, with specific "tropism" of the virus to certain target tissues only. In experimental infections this course of development has been observed repeatedly, but the underlying mechanisms can only be speculated on (*Ebert* and *Wilt*, 1960). Our study was prompted by an example from clinical observations on sensitive periods, the well-known fact that maternal rubella infection causes congenital cataract during the first six weeks of gestation only (*Töndury* and *Smith*, 1966). In terms of lens formation, this time corresponds to the early stages of induction, with invagination of the surface ectoderm to form a lens vesicle which subsequently detaches from the outer body surface. This period also parallels various developmental events in fetomaternal relations during early pregnancy. Knowing, however, that susceptibility to the same virus extends for longer periods of time in the developing inner ear and heart, we were inclined to search for the answer to the restriction of sensitivity at the organ level.

As a model for study we first employed chick embryos.
After experimental infection with mumps virus, they were reported to develop cataracts similar to those caused by rubella virus in human embryos (*Williamson* et al., 1957; *Robertson* et al., 1964). According to these observations, the developing lens is vulnerable to the viral infection at the open lens-vesicle stage only. The authors suggested that if it is the

virus itself – not an indirect effect – that causes cataract, it may gain entrance to the open lens vesicle from the exterior and become trapped there on closure of the vesicle. With no escape it might then continue replication in this "incubation chamber," building up high titers and eventually destroying the lens (*Robertson* et al., 1964). We were challenged by the idea that the lens, being an avascular organ and thus beyond the reach of systemic protective mechanisms, should have an built-in mechanism limiting the teratogenic action of the virus from a certain stage onward. This event should, therefore, be demonstrable in vitro.

b) Results

We started out with a large series of experiments in ovo, confirming the above-mentioned results of *Williamson* et al. (1957), and *Robertson* et al. (1964). In optimal conditions cataracts were produced in ca. 50% of the embryos infected in the open lens vesicle stage, but never if infected in the closed stage (*Karkinen-Jääskeläinen*, 1973). Histologic criteria proved insufficient for localization of the replicating virus and we employed the techniques of immunofluorescence to follow the course of infection. Embryos at the closed stage never showed any viral antigens within the lenses. In them, virus replication took place in the cells of the outer body surface ectoderm only. The younger embryos exhibited specific fluorescence inside the cataractous lenses and in the neural ectoderm as well as in the surface ectoderm. The presence of the virus in the cataractous lenses was further confirmed by electron microscopy, which revealed intense accumulation of viral nucleoproteins in the cytoplasm of the lens fibers. A rather low output of mature viruses was observed, suggesting an imbalanced replication which might contribute to the ultimate destruction of the fibers (*Virtanen* et al., 1973).

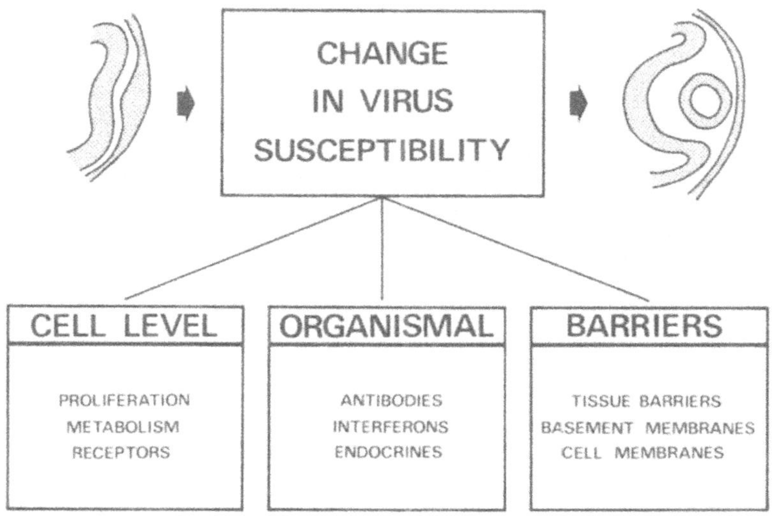

Fig. 6. Scheme of three possible levels at which susceptibility to virus may be restricted: cellular, organismal, and morphogenetic

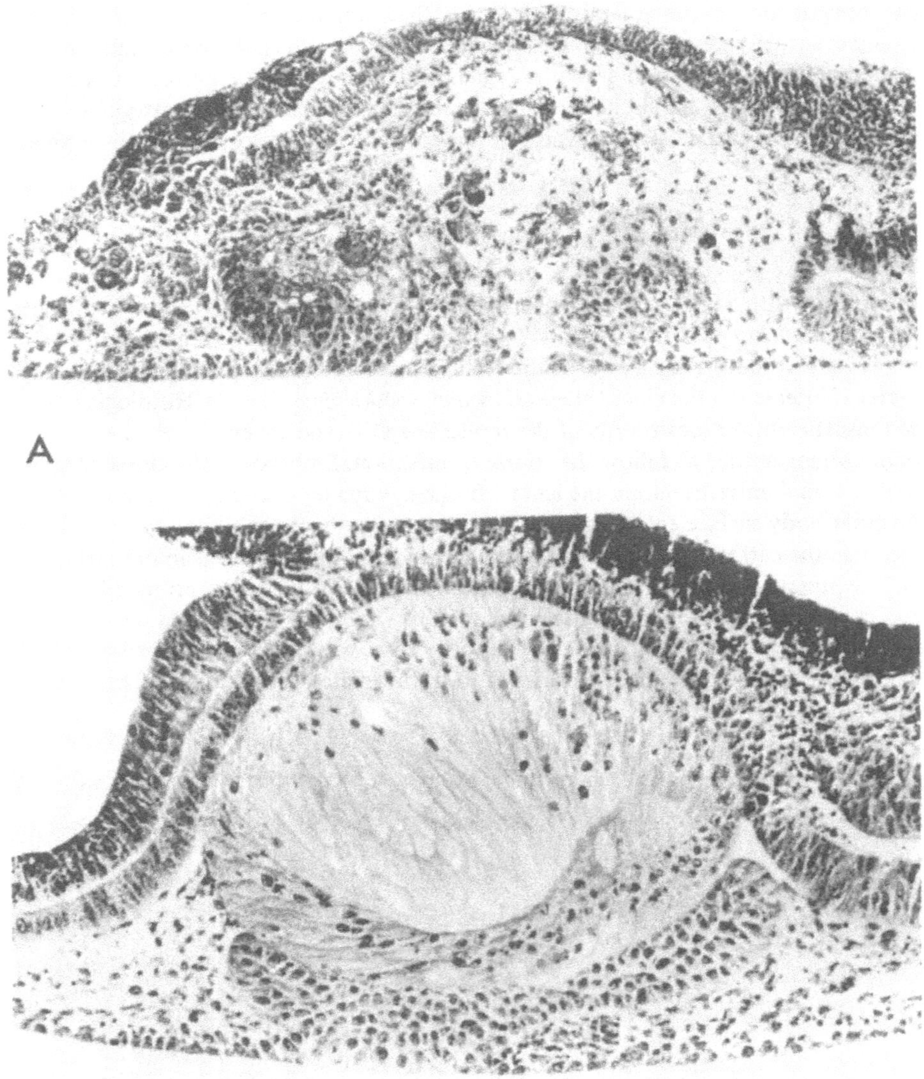

Fig. 7. Eye rudiment of chick embryo taken before closure of lens vesicle and grown in vitro for five days. Infected explant showing destruction of fibers (A), not detectable in uninfected control (B). (*Karkinen-Jääskeläinen*, 1973)

Results of the in ovo experiments on chick embryos excluded the involvement of any maternal factor in the restriction of the sensitive period (Fig. 6) and confirmed the presence of virus in the cataractous lenses. To exclude other organismal factors (Fig. 6) we then experimented with eye rudiments in organ culture (*Karkinen-Jääskeläinen*, op. cit.). This procedure offered one additional advantage: It made possible individual and exact timing of infection of the rudiment in relation to its morphological stage of development. It became evident that the lens remained sensitive to the infection until closure of the lens vesicle. Infection before that event resulted in severe cell damage during subsequent fiber differentiation (Fig. 7A). Inside such lenses viral antigens were always detectable by immunofluorescence. On the contrary, rudiments infected a few hours later, when closure of the lens vesicle was complete, continued to differentiate quite normally. They were indistinguishable from uninfected controls (Fig. 7B) and showed no viral antigens within the lens (Fig. 9).

The results of the in vitro experiments thus confirmed those of the in ovo series and excluded organismal factors, such as an immune response or endocrine effects, as restrictive mechanisms. Cellular synthesis of interferons was not entirely excluded, but is made highly unlikely as a decisive factor by the fact that loss of sensitivity to virus took place in a few hours and was clearly correlated with a morphologic change. Two possible explanations for the restriction of susceptibility to the virus now remained (Fig. 6). First, the outer body surface epithelium might function as a barrier, or the basement membrane surrounding the lens and thickening to a multilaminated lens capsule at that time could prevent penetration of virus particles into the lens vesicle. Second, maturation of the differentiating cells themselves might result in insensitivity to the virus. Examples of such a development of resistance have been reported (see *Saxén* and *Rapola*, 1969). In our laboratory the mouse kidney rudiment was found to undergo a drastic change in susceptibility to the virus during its development in vitro. Undifferentiated mesenchymal cells were highly susceptible to polyoma virus, but differentiation into pretubular condensates paralleled loss of susceptibility to infection (*Saxén* et al., 1962; *Vainio* et al., 1963). This happened before any sign of basement membrane formation was evident (*Wartiovaara* et al., 1965). This "maturation resistance" could be brought about by various mechanisms, such as changes in the proliferation rate or metabolism of the cells or loss of the machinery necessary for virus replication. Changes at the cell surface with disappearance of receptor sites for the virus are also possible. To distinguish between changes at cell and tissue level we now continued experiments with the closed-stage rudiments. The suspected barriers were experimentally by-passed (*Karkinen-Jääskeläinen*, op. cit.) by making a cut through the outer surface epithelium with a knife before infection. The wound was kept open with a slip of filter membrane to allow viral entrance. The operation was then repeated through both the epithelium and the lens capsule. Cutting of the epithelium alone did not have any detectable effect. A wound through the lens capsule had a profound effect, however, with destruction of the lens fibers of an infected rudiment during subsequent culture in vitro (Fig. 8A). In an uninfected control some lens material was extruded through the wound, but no cell damage ensued (Fig. 8B). In cases where an infected rudiment had an intact capsule no viral antigens were detected inside the lens (Fig. 9), but the spread of those antigens into the wounded and damaged lenses was clearly demonstrated by immunofluorescence (Fig. 10).

We then returned to our original question relating to the human embryonic lens with limited stage of susceptibility to rubella virus. An identical restrictive mechanism was found to operate. Human embryonic eye rudiments differentiate well in vitro and grow for long

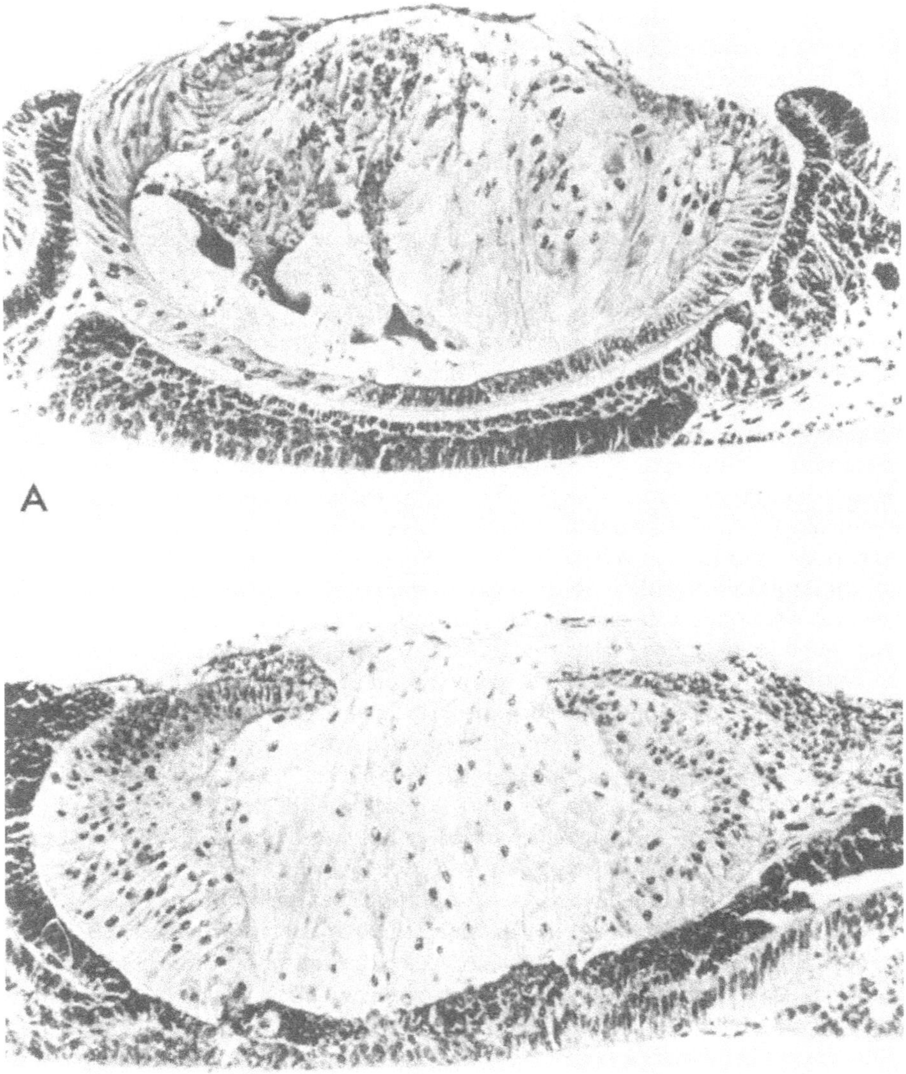

Fig. 8. Chick eye rudiments after lens vesicle had closed. Cut was made through outer surface of ectoderm and lens capsule. In infected explant profound cell destruction ensued during subsequent cultivation (A). In uninfected control extrusion of lens material through wound was seen but no cell damage (B). (*Karkinen-Jääskeläinen*, 1973)

periods of time in organ cultures (Fig. 11A). If such rudiments are infected with rubella virus in the open lens-vesicle stage or after a cut has been made through the capsule before

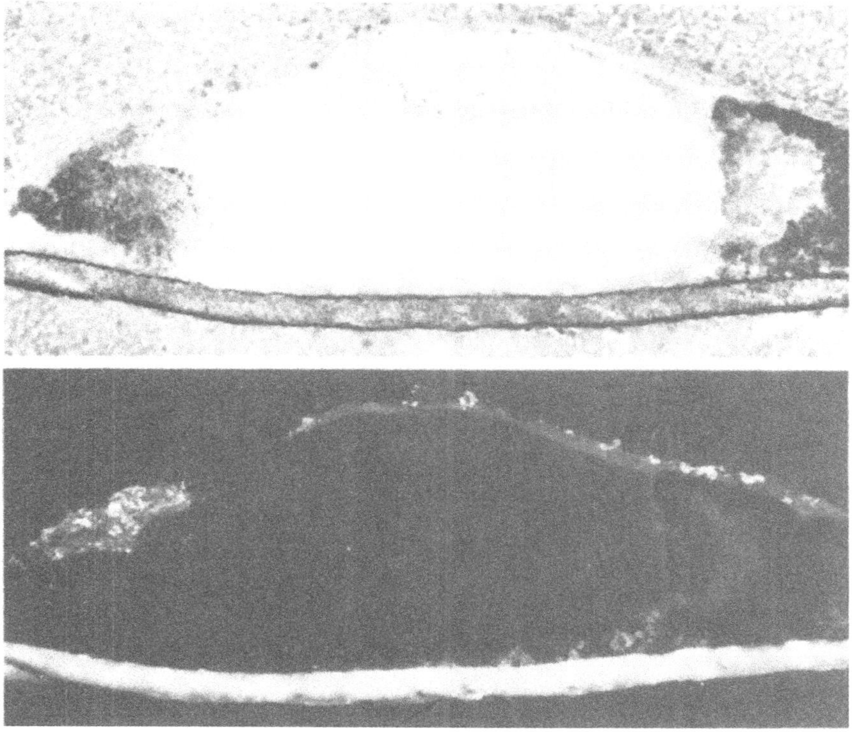

Fig. 9. Immunofluorescence revealed no viral antigens inside embryonic chick lens with intact capsule. Specific fluorescence is seen in surface ectoderm

infection, destruction of the lens fibers is seen during subsequent development in vitro (Fig. 11B). Cell damage is accompanied by the appearance of viral antigens inside the damaged lenses (*Karkinen-Jääskeläinen* et al., 1975). The human embryonic lens, however, has to be cultured for a much longer time than chick lens before any cell damage is detectable.

c) Conclusions

We concluded that here we are dealing with a relatively simple mechanism that restricts susceptibility to virus. The lens cells themselves remain susceptible in spite of their differentiation, but during normal development the embryo develops a barrier which protects the susceptible cells from penetration of the virus (*Karkinen-Jääskeläinen*, op. cit.).

Evidently this example is but one of several mechanisms restricting viral teratogenicity. The progressive limitation of susceptibility in an embryo, from early destruction, to strict compartmentalization of damage, and finally to specific "trophism" of the virus to target organs in the maturing organism, is without doubt due to multifactorial regulation. Yet, in view of the considerable overlap of the narrowly limited sensitive periods of various developing organs, the solution to the problem might be sought for at organ level. Whether it proves to be a "maturation resistance" of the cells in a given stage or a morphogenetic event, as above, it could apply to other teratogens as well, on occasions.

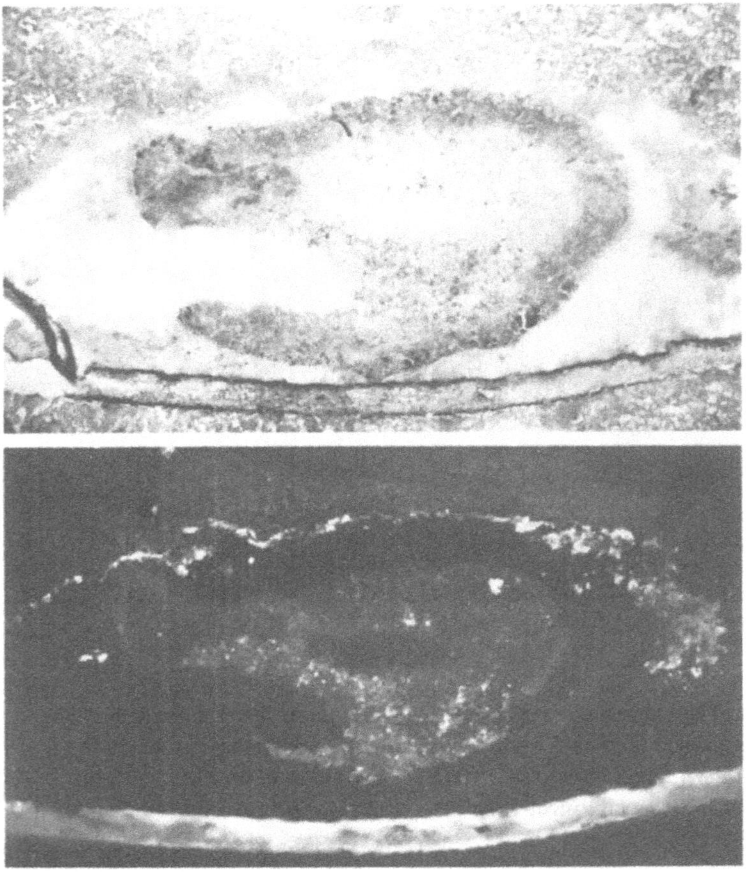

Fig. 10. Viral antigens detectable by immunofluorescence within a chick embryo lens infected through cut in protective barriers

3. Interaction of Genetic and Environmental Factors

a) Introduction

Cleft palate is a typical example of a malformation with multifactorial etiology, where genetic and environmental factors interact. In experimental animals cleft palate has been produced by several teratogens, of which corticoids have been the most widely used

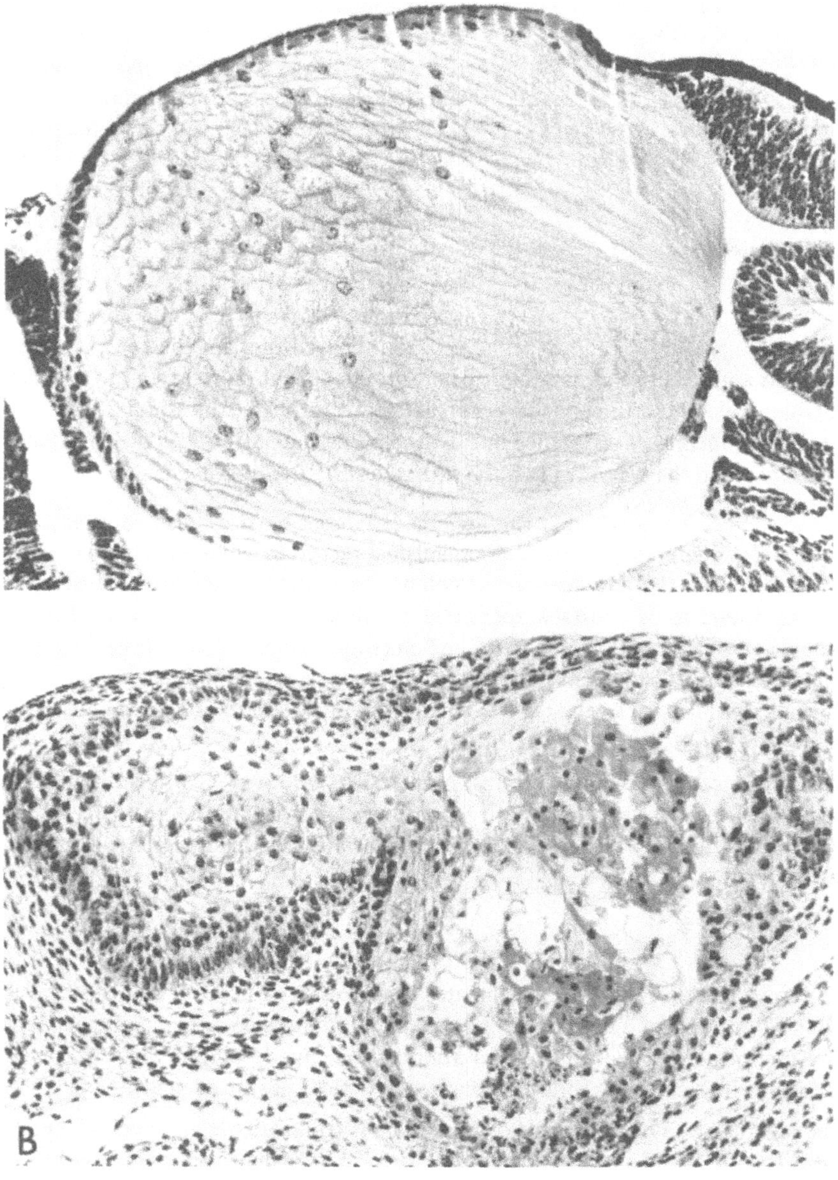

Fig. 11. Human embryo lens in vitro: After three week's cultivation somewhat slow but good differentiation is seen (A). If infection occurs in open lens vesicle stage, penetration of rubella virus is followed by damage to lens fibers (B)

(*Fraser* and *Fainstat*, 1951). The interaction of genetic and environmental factors has been clearly demonstrated in experiments with inbred strains of mice (*Kalter*, 1965). The susceptibility to cortisone-induced cleft palate is distinctly influenced by the genotype (Table 3).

Table 3. Frequency of cleft palate in offspring of cortisone-treated mothers of different inbred strains of mice (*Kalter*, 1965)

Strain	Percent
CBA	12
C57BL	19
C3H	68
DBA	92
A	100

It was found that closure of palatal shelves occurred at a later stage of development in the strains that were sensitive to cortisone treatment than in the resistant strain, and that cortisone treatment delayed initiation of the shelf movement and slowed down the process of closure (*Walker* and *Fraser*, 1956, 1957). Later *Fraser* (1961, 1965, 1969) presented a hypothesis to explain these strain differences. He postulated that there is a developmental threshold, i.e., the relation of the growth of the head and oral cavity to the shelf movement, beyond which the palatal shelves can no longer fuse. Thus, if cortisone caused the same degree of retardation in the resistant and sensitive strains, cleft palate would appear more often in the sensitive strain, in which palatal closure already occurs later. Accordingly, the strain difference would be due to an "intrinsic" difference in the developmental patterns of the embryos, rather than a difference in the way in which the strains handle the teratogen.

We wanted to know whether this genetic difference between a susceptible and a resistant strain was due merely to such an organismal factor, or whether there was a difference in the response to the teratogen at the cell or tissue level as well. We used organ culture, previously found to be a suitable technique for study of the development of the secondary palate (*Moriarty* et al., 1963; *Lahti* and *Saxén*, 1967; *Pourtois*, 1968). In these conditions the various organismal factors, as mentioned above, are eliminated, and in addition the developmental stage of the palates when exposed to the teratogen could be exactly determined.

We used mice of two inbred strains, A and CBA, whose in vivo sensitivity to hydrocortisone had been tested earlier in our laboratory: Hydrocortisone, injected on days 11-14 of pregnancy caused 56% cleft palate in strain A but none in strain CBA (*Lahti* et al., 1972).

b) Results

As can be seen in Figure 12, the palates developed well in vitro, and the shelves fused after 3 or 4 days of cultivation. The developmental stage of the palates at the start of cultivation influenced the time required for closure, the developmentally younger palates closing more slowly than those explanted at a more advanced stage.

The palates of 13-day embryos of the resistant strain CBA were more advanced in development than those of strain A, and thus a search was made for developmentally comparable palates of both strains. The developmental stage of strain A palates resembled that of 13– day palates of strain CBA if the former were harvested 6 h later (13+). Correspondingly, CBA palates had to be collected 7 h earlier if they were to be caught in the same developmental stage as 13– day strain A palates. The development of these comparable groups of palates was then followed from camera lucida drawings made at 6 h intervals until no open space could be seen between the shelves.

Hydrocortisone in a concentration of 1 μg/ml caused prolongation of the palatal closure time of strain A explants taken at both 13 and 13+ days (Figs. 13A and B). On the con-

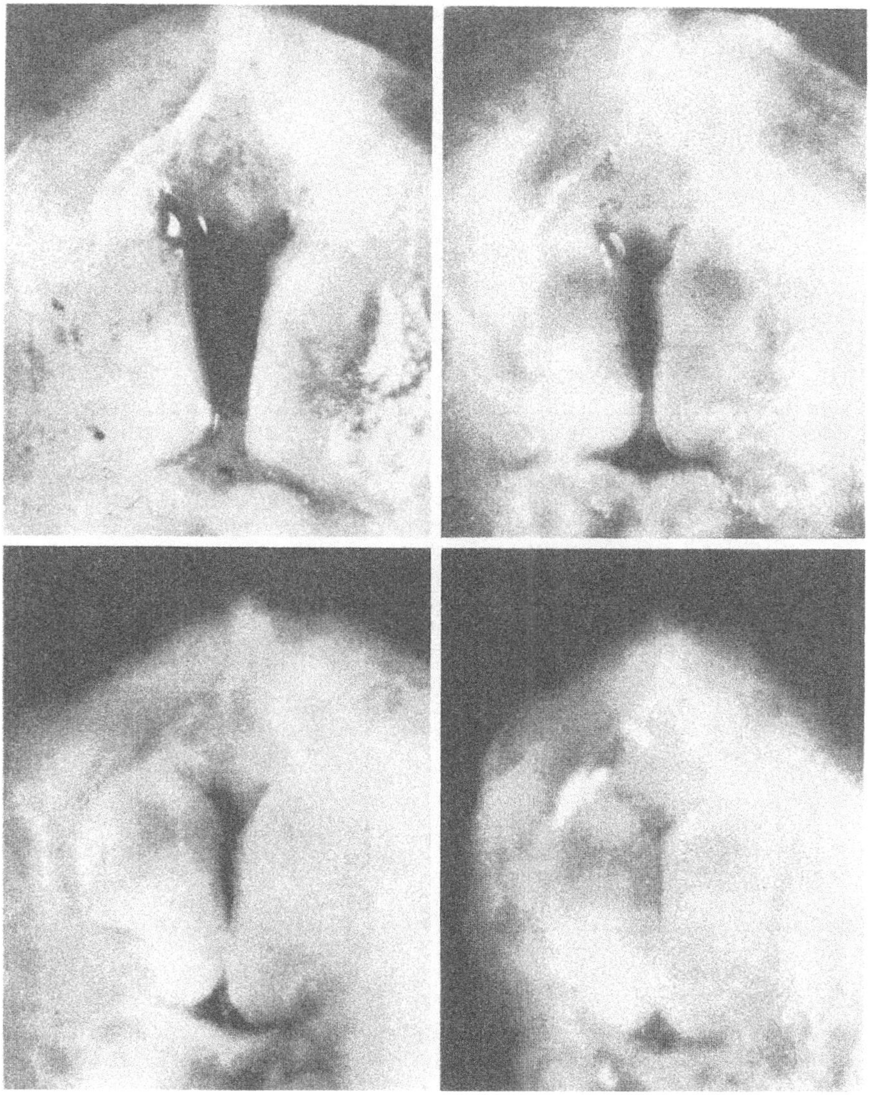

Fig. 12. Development of secondary palate of 13-day mouse embryo in vitro, photographed at 24-h intervals (*Lahti* et al., 1972)

trary, the palates of strain CBA at corresponding developmental stages (13– and 13 respectively) failed to show any response to hydrocortisone treatment (Figs. 13C and D).

The experiment was also performed on hybrids CBA ♀ × A ♂ at 13 days. In palatal stage these hybrids resemble strain CBA more than strain A and yet palatal closure in vitro was slightly prolonged by hydrocortisone. The response of the hybrids was thus intermediate between those of the parental strains.

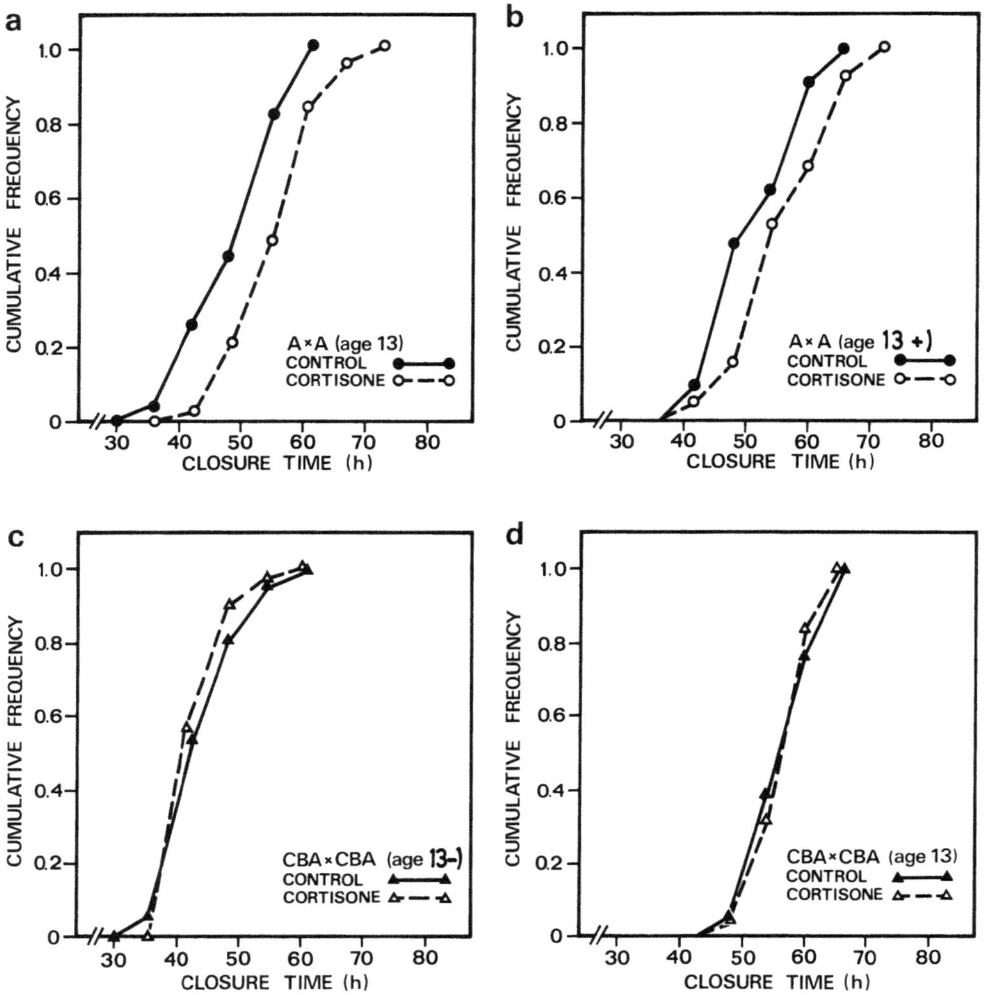

Fig. 13. Effect of hydrocortisone on in vitro closure of palatal shelves in embryos of strains A and CBA at two corresponding stages of development. Cumulative index: Fraction of closed palates at 6-h intervals. (After *I. Saxén*, 1973)

c) Conclusions

Thus, in vitro there was a definite difference in response to hydrocortisone treatment between the strains that differed in susceptibility to hydrocortisone-induced cleft palate in vivo. This difference should be found somewhere at tissue level. At present we are trying to find an explanation for this difference.

There is no doubt that the developmental stage of the palate in relation to that of the embryo is of importance in determining the predisposition of the strain to cleft palate, as originally pointed out by *Fraser*. His hypothesis is actually supported by our observation that the final fusion of palatal shelves was never prevented by hydrocortisone in organ culture, even though the time of closure was prolonged. Our experiments eliminated the effect of the growth of the surrounding tissues, which was assumed to create the developmental threshold. We suggest, however, that the genetically determined differential rate of development is not necessarily the only mechanism governing the differences between these strains. Our observation of a difference in the response to hydrocortisone between the two strains even in vitro leads us to conclude that the metabolism and mode of action of a teratogen at cell and tissue level also influences the susceptibility to cleft palate.

III. Concluding Comments

The three examples of in vitro studies presented above suggest that the organ culture technique can be used in the field of experimental teratology. Especially when defined targets and relatively short periods of development are analyzed, the technique offers many advantages. On the other hand, the critical reader has also realized the most obvious drawback of this technique: Single and relatively simple events of embryogenesis are dissociated from their normal involvement in the synchronized development of the organism and the observations may not be applicable to the corresponding situations in vivo. Consequently, although the organ culture technique should be considered a valuable tool for a reductionistic approach in studies of abnormal development, it is merely a complementary method and should never be regarded as a substitute for epidemiologic studies or experiments in vivo.

References

Bevelander, G., Nakahara, H., Rolle, G.K.: The effect of tetracycline on the development of the skeletal system of the chick embryo. Develop. Biol. 2, 298-312 (1960).

Bevelander, G., Rolle, G.K., Cohlan, S.Q.: The effect of the administration of tetracycline on the development of teeth. J. dent. Res. 40, 1020-1024 (1961).

Biggers, J.D., Gwatkin, R.B.B., Heyner, S.: The growth of avian and mammalian tibiae on a relatively simple chemically defined medium. Exp. Cell Res. 25, 41-58 (1961).

Charles, D.: Placental transmission of antibiotics. J. Obstet. Gynaec. Brit. Emp. 61, 750-757 (1954).

Cohlan, S.Q., Bevelander, G., Tiamsic, T.: Growth inhibition of prematures receiving tetracycline. Amer. J. Dis. Child. 105, 453-461 (1963).

Ebert, J.D., Wilt, F.H.: Animal viruses and embryos. Quart. Rev. Biol. 35, 261-312 (1960).

Fell, H.B.: Techniques of bone cultivation. In: Methods in Medical Research, Vol. IV (ed. *M.B. Visscher*), p. 234. Chicago: Year Book Medical Publishers Inc. 1951.

Fell, H.B.: Changes in synthesis induced in organ cultures. In: Synthesis of Molecular and Cellular Structure (ed. *D. Rudnick*), pp. 139-160. New York: Ronald Press Co. 1961.

Fraser, F.C.: The use of teratogens in the analysis of abnormal developmental mechanisms. In: First International Conference on Congenital Malformations (ed. *M. Fishbein*), pp. 179-186. Philadelphia: Lippincott 1961.

Fraser, F.C.: Experimental teratogenesis in relation to congenital malformations in man. In: Second International Conference on Congenital Malformations (ed. *M. Fishbein*), pp. 277-287. New York: The International Medical Congress Ltd. 1964.

Fraser, F.C.: Some genetic aspects of teratology. In: Teratology: Principles and Techniques (eds. *J.G. Wilson, J. Warkany*), pp. 21-28. Chicago: University of Chicago Press 1965.

Fraser, F.C.: Gene-environmental interactions in the production of cleft palate. In: Methods for teratological studies in experimental animals and man (eds. *H. Nishimura, J.R. Miller*), pp. 34-49 Tokyo: Igaku Shoin 1969.

Fraser, F.C., Fainstat, T.D.: Production of congenital defects in the offspring of pregnant mice treated with cortisone. Pediatrics 8, 527-533 (1951).

Halme, J., Kivirikko, K.I., Kaitila, I., Saxén, L.: Effect of tetracycline on collagen biosynthesis in cultured embryonic bones. Biochem. Pharmacol. 18, 827-836 (1969).

Kaitila, I.: Effect of tetracycline in mineralization in cycloheximide-treated bones in vitro. Calcif. Tiss. Res. 7, 46-57 (1971a).

Kaitila, I.: The mechanism by which tetracycline hydrochloride inhibits mineralization in vitro. Biochim. biophys. Acta (Amst.) 244, 584-594 (1971b).

Kaitila, I., Wartiovaara, J., Laitinen, O., Saxén, L.: The inhibitory effect of tetracycline on osteogenesis in organ culture. J. Embryol. exp. Morph. 23, 185-211 (1970).

Kalter, H.: Interplay of intrinsic and extrinsic factors. In: Teratology: Principles and Techniques (eds. *J.G. Wilson, J. Warkany*), pp. 57-80. Chicago: University of Chicago Press 1965.

Karkinen-Jääskeläinen, M.: Spatial and temporal restriction of mumps virus induced lesions in the developing chick lens. Acata path. microbiol. scand., Sect. A, Suppl. 243, 1-52 (1973).

Karkinen-Jääskeläinen, M., Saxén, L., Vaheri, A., Leinikki, P.: Rubella cataract in vitro: Sensitive period of the developing human lens. J. exp. Med. 141, 1238-1248 (1975).

Katz, E.P.: The kinetics of mineralization in vitro. I. The nucleation properties of 640-Å collagen at 25°. Biochim. biophys. Acta (Amst.) 194, 121-129 (1969).

Lahti, A., Antila, E., Saxén, L.: The effect of hydrocortisone on the closure of the palatal shelves in two inbred strains of mice in vivo and in vitro. Teratology 6, 37-42 (1972).

Lahti, A., Saxén, L.: Effect of hydrocortisone on the closure of palatal shelves in vivo and in vitro. Nature (Lond.) 216, 1217-1218 (1967).

Lash, J.W., Saxén, L.: Human teratogenesis: In vitro studies on thalidomide inhibited chondrogenesis. Develop. Biol. 28, 61-70 (1972).

Milch, R.A., Rall, D.P., Tobie, J.E.: Fluorescence of tetracycline antibiotics in bone. J. Bone Jt. Surg. 40A, 897-910 (1958).

Moriarty, T.M., Weinstein, S., Gibson, R.S.: The development in vitro and in vivo of fusion of the palatal processes of rat embryos. J. Embryol. exp. Morph. 11, 605-619 (1963).

Pourtois, M.: La fusion des cretes palatines et son altération par quelques agents tératogènes. Arch. Biol. (Liège) 79, 105-179 (1968).

Robertson, G.G., Williamson, A.P., Blattner, R.J.: Origin and development of lens cataracts in mumps-infected chick embryos. Amer. J. Anat. 115, 473-486 (1964).

Saxén, I.: Effect of hydrocortisone on the development in vitro of the secondary palate in two inbred strains of mice. Arch. oral Biol. 18, 1469-1479 (1973).

Saxén, L.: Tetracycline: Effect on osteogenesis in vitro. Science 149, 870-872 (1965).

Saxén, L.: Effect of tetracycline on osteogenesis in vitro. J. exp. Zool. 162, 269-294 (1966).

Saxén, L.: Population surveillance for birth defects. In: Birth defects (eds. *A.G.Motulsky, W. Lenz*), pp. 177-186. Amsterdam: Excerpta Medica Foundation 1974.

Saxén, L., Kaitila, I.: The effect and mode of action of tetracycline on bone development in vitro. Advanc. exp. Med. 27, 205-218 (1972).

Saxén, L., Koskimies, O., Lahti, A., Miettinen, H., Rapola, J., Wartiovaara, J.: Differentiation of kidney mesenchyme in an experimental model system. Advanc. Morphogenes. 7, 251-293 (1968).

Saxén, L., Rapola, J.: Congenital defects. New York: Holt, Rinehart & Winston 1969.

Saxén, L., Vainio, T., Toivonen, S.: Effect of polyoma virus on mouse kidney rudiment in vitro. J. nat. Cancer Inst. 29, 597-631 (1962).

Simpson, D.M., Burnette, J.C., Bawden, J.W.: Maternal-fetal blood tetracycline levels in guinea pigs. J. oral. Therap. Pharmacol. 3, 403-408 (1967).

Strangeways, T.S.P., Fell, H.P.: Experimental studies on the differentiation of embryonic tissues growing in vivo and in vitro. I. The development of the undifferentiated limb-bud (a) when subsequently grafted into the postembryonic chick and (b) when cultivated in vitro. Proc. roy. Soc. B 99, 340-364 (1926).

Töndury, G., Smith, D.W.: Fetal rubella pathology. J. Pediat. 68, 867-879 (1966).

Trowell, O.A.: Problems in the maintenance of mature organs in vitro. In: La Culture Organotypique (ed. *E. Wolff*), pp. 237-249. Paris: C.N.R.S. 1961.

Vainio, T., Saxén, L., Toivonen, S.: The acquisition of cellular resistance to polyoma virus during embryonic differentiation. Virology 20, 380-385 (1963).

Virtanen, I., Karkinen-Jääskeläinen, M., Saxén, L.: Ultrastructural characteristics of mumps virus replication in the cataracts of chick embryonic lens. Exp. Eye Res. 17, 463-472 (1973).

Walker, B.E., Fraser, F.C.: Closure of the secondary palate in three strains of mice. J. Embryol. exp. Morph. 4, 176-189 (1956).

Walker, B.E., Fraser, F.C.: The embryology of cortisone-induced cleft palate. J. Embryol. exp. Morph. 5, 201-209 (1957).

Wallman, I.S., Hilton, H.B.: Teeth pigmented by tetracycline. Lancet 1962 I, 827-829.

Wartiovaara, J., Saxén, L., Vainio, T.: Viral susceptibility and embryonic differentiation. V. Localization of virus-like particles in mouse kidney rudiment infected with polyoma virus. Acta path. microbiol. scand. 63, 72-78 (1965).

Wessells, N.K.: Avian and mammalian organ culture. In: Methods in Developmental Biology (eds. *F.H. Wilt, N.K. Wessells*), pp. 445-456. New York: Thomas Y. Crowell Company 1967.

Williamson, A.P., Blattner, R.J., Simonsen, L.: Cataracts following mumps virus in early chick embryos. Proc. Soc. exp. Biol. (N.Y.) 96, 224-228 (1957).

Teratogenic Effects and Placental Permeability of Heavy Metals

V.H. FERM

I. Introduction

Embryonic failure may result from a variety of environmental influences. Certainly experimental mammalian teratogenesis has demonstrated that a broad spectrum of stimuli can induce a variety of morphological developmental defects in experimental animals. There are clearly identifiable environmental teratogenic stimuli, such as rubella, drugs, radiation, etc., which deleteriously affect human embryonic development.

We have been investigating the teratogenic properties of heavy metals in mammalian systems. Some of these metals are potentially serious pollutants in industrialized societies and as such might play important roles in human disease. Little is known concerning the possible effect of heavy metals on mammalian embryogenesis.

The heavy metals offer a number of advantages as investigative tools for experimental teratology. They enter into a number of chemical reactions within the living organism especially as important components of the metalloenzymes. Several sophisticated techniques, including radioautography, isotope counting, and electron probe analysis, are available for definitive quantitative and site localization studies.

In any mammalian teratological study, the effect of a teratogen might be exerted by an effect on the maternal system indirectly then affecting the embryo, by alteration of placental function or permeability, or by a direct effect of the teratogen on embryonic tissues. Again, the heavy metals offer the opportunity for a precise methodology in separating out these possible mechanisms.

II. Animal Model

We have used the pregnant golden hamster *(Cricetus auratus)* as the experimental animal in our testing system. This animal breeds readily and it is easy to obtain accurately-timed matings. The average litter size is 12 pups and the embryo responds to a variety of well-known teratogenic agents. Because of the extremely short gestation period in this animal (16 days) embryogenesis proceeds at a fast rate. Figure 1 shows the rapid progression of development between the eighth and ninth day of gestation. The major stages of organogenesis occur during this 24 h period. Thus, in our system the teratogenic stimulus is introduced at a single selected time during this period, and the embryos or fetuses collected at specified times thereafter.

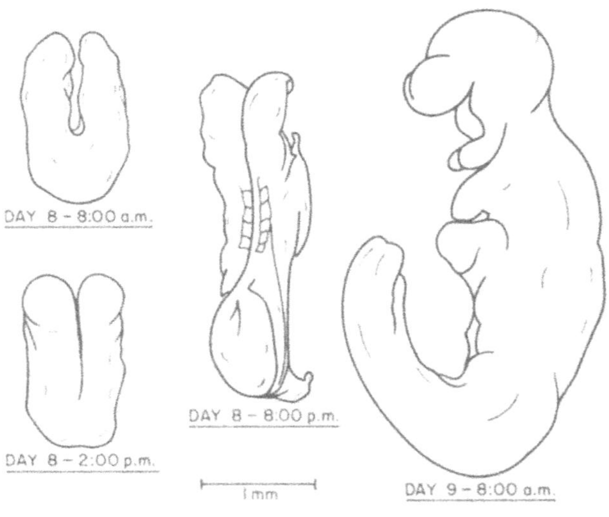

Fig. 1. Diagrammatic representation of embryonic development in golden hamster from day 8-9 of intrauterine life. Development from primitive streak to early limb bud stage occurs during this period

The intravenous injection of various heavy metals and combinations thereof, preclude the variable blood levels which might follow oral or intraperitoneal injection. We inject all solutions into the lightly anesthetized pregnant hamster via the lingual vein on the morning of the eighth day of gestation when the embryo is in the early primitive streak stage (Fig. 1). Embryos and other tissues are then collected, either 24 h later (day 9) or 96 h later (day 12). A record is made of the number of embryonic resorptions that have occurred as a result of the teratogenic treatment and the surviving embryos or fetuses are examined for gross external developmental malformations. Randomly selected fetuses are fixed in alcohol and the skeletons subsequently stained with alizarin red. This latter technique is used to determine possible skeletal defects.

In addition, we have studied the possibility that the introduction of the teratogen at different times during the period of organogenesis may induce a spectrum of malformations peculiar to the specific metal. In such instances, the metal has been introduced into the pregnant animal as a single injection at 6 h intervals during the organogenetic period. The resulting abnormalities have been identified as the "teratogenic profile" for the metal concerned.

III. Specific Teratogenic Metals

1. Lead

This metal induces rather specific effects on the tail buds of hamster embryos when injected into pregnant hamsters. As shown in Figure 2, treatment of mothers at other times during the major critical organogenetic periods does not change the spectrum of malformations in the offspring (*Ferm* and *Ferm*, 1971).

	BRAIN	PALATE-FACE	HEART	FORELIMB	HINDLIMB	RIBS	URO-GENITAL	SACRUM-TAIL
Pb								++++
Cd		++++	+	++++	++++			
As	++++					++++	++++	
Cu			++					
In				++++*	++++*			

*-digits only

Fig. 2. Teratogenic profile of several heavy metals in golden hamster. Single injections of metals at 6 h intervals resulted in spectrum of malformations noted in chart. Some metals (Cd, As) demonstrated range of defects, while others (Pb, In) are rather specific in their effect

The peculiar tail bud lesion is first noted histologically on the day 9 of gestation as an edema of the tissues surrounding the caudal neural tube. Large fluid-filled spaces or "blisters" then appear, followed by hemorrhage into these spaces. Healing follows, but there is considerable distortion of the caudal neural tube, notochord and sacral vertebrae (*Carpenter* and *Ferm*, 1974). The specificity of this effect suggests a possible comparison to certain tail-affected mutants (phenocopies) as described in mice (*Grüneberg*, 1963).

2. Cadmium

The single injection of cadmium into pregnant hamsters at one of the 6 h intervals during the 24 h critical period of embryonic organogenesis results in a particular spectrum of malformations (*Ferm*, 1971). This spectrum is shown in Figure 2. Administration of cadmium during the early part of this susceptible period (day 8–8a.m.) results in severe facial clefts, especially of the lip and palate. Later, (day 8–8p.m.), the forelimbs are affected to degrees ranging from absence of a single digit to complete amelia. The spectrum shifts to lower limb abnormalities when the cadmium is administered during the late stages of the susceptible period (day 9–8a.m.).

It is now well-known that zinc and cadmium interact or complete in a variety of biological systems. In our animal model we have found that zinc administered to pregnant hamsters is not teratogenic (*Ferm* and *Carpenter*, 1968) although *Hurley* and *Swenerton* (1966) have shown that a zinc deficient diet does cause malformations in rats. The administration of zinc, however, to cadmium treated animals protects the embryo rather strikingly and the incidence of malformations as well as the embryonic resorption rate is remarkably reduced. The zinc must be administered within a relatively short period of time following or preceding the cadmium insult. Selenium also has a protective effect against the malformations induced by cadmium in the hamster embryo (*Holmberg* and *Ferm*, 1969).

3. Arsenic

Utilizing the technique as previously described, we have been able to induce a spectrum of effects with sodium arsenate (*Ferm* et al., 1971). Many of these malformations are peculiar, at least in this animal model system, to arsenic. For example, this metal has specific and severe effects on the genito-urinary system in both male and female embryos. The range of induced malformations is from hypoplasia to complete agenesis of the uterus, tubes, testis, and kidneys. Other less specific malformations include exencephaly and rib defects. The cleft lip and palates seen in cadmium treatment and the tail malformations peculiar to lead are not found in the arsenic-treated animals. It has been reported that selenium has a significant protective effect against the congenital anomalies induced by arsenic (*Holmberg* and *Ferm*, 1969).

4. Copper

The role of copper in embryonic development is not at all clear. Copper deficiency causes an increase in embryonic deaths in both the chick and the rat (*Dutt* and *Mills*, 1960; *O'Dell*, 1968). Embryos surviving such treatment, in both species, reveal developmental malformations. Excess copper presumably from the composition of intrauterine loops prevent implantation and blastocyst development in the rat, hamster, and rabbit (*Chang* et al., 1970; *Zipper* et al., 1970). *Brinster* and *Cross* (1972) have shown that copper is toxic to preimplantation mouse embryos maintained in tissue culture media. Our studies have shown that copper in a chelated form (citrate) was considerably more embryopathic than copper in the uncomplexed (sulfate) form (*Ferm* and *Hanlon*, 1974a). In our animal model, malformations of the heart (especially ectopia cordis) appeared to be a specific teratogenic effect of copper, more so with the citrate complex. This may be the result of the greater binding of uncomplexed copper to sites in the maternal system and thus its relative unavailability to the developing embryo.

5. Indium

In screening other metals for teratogenic effects in this animal model, indium produced some minor but specific teratogenic effects. The target organ for this metal was the distal portion of the extremity, causing fusion, stunting, and even polydactyly in the digits (*Ferm* and *Carpenter*, 1970). All limbs appeared to be affected. Indium appears to be a rather potent teratogen (Fig. 3) but the type of malformation induced by this metal is relatively minor (Fig. 2).

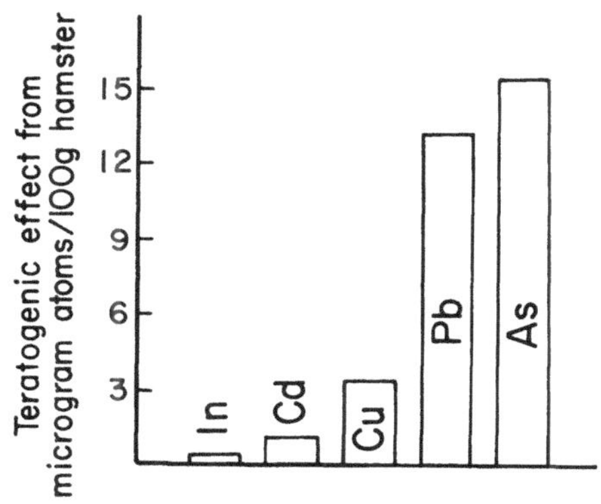

Fig. 3. Comparison of activity of 5 heavy metals in golden hamster based on number of microgram atoms of metal which induce its teratogenic effect

IV. Placental Permeability to Heavy Metals

One possible mechanism of teratogenic action of the heavy metals on embryonic system is the possibility of a direct action on embryonic tissues. One hypothesis would be that because some of these metals are important components of many metalloenzymes, an imbalance (excess or deficiency) would result in a competition between metals for active sites in such important enzymes. The resulting disturbance in the function of such enzymes may be the causal mechanism of the malformation.

We now have evidence for the placental permeability of three of the teratogenic metals during the critical stages of embryogenesis in our model system. Studies on the permeability of the placenta to radioactive lead (Pb^{209}) by radioautographic techniques have shown that lead crosses the placenta of the hamster embryo during this critical period (day 8-9). While there does not appear to be any preferential localization of lead within specific embryonic tissues, the yolk sac placenta at this stage has a considerable concentration of lead (*Carpenter* et al., 1973).

Cadmium penetrates the placenta during the critical stages of embryogenesis and the simultaneous administration of zinc, even though zinc protects against the teratogenic effect of cadmium (*Ferm* et al., 1970), does not affect this permeability. Recent studies have shown that the placental transfer of radioactive zinc under identical experimental conditions is not affected by the presence of excess cadmium (*Ferm* and *Hanlon*, 1974b).

Similar studies on the permeability of the copper ion have indicated that, at least in the form of copper citrate, this metal also penetrates the placenta and can be detected in the tissues of the day 9 hamster embryo (*Ferm* and *Hanlon*, 1974a).

V. Conclusion

Five heavy metals have been shown to have rather specific teratogenic profiles in the golden hamster. Figure 3 shows that the sensitivity based on the teratogenic activity of microgram atoms per animal varies from metal to metal as does the severity of the teratogenic response (Fig. 2). Data for three of these metals (lead, cadmium, and copper) reveal that they cross the placental membrane during the critical stages of embryogenesis and can be found in embryonic tissue.

Each of these metals cause a specific syndrome or pattern of developmental malformations. Some of these syndromes are not unlike well-known genetic phenocopies which have been clearly established in other species.

Because several of these metals are known to be important components of many metalloenzymes, one explanation for their teratogenic activity may well be that they interfere with metalloenzymes during critical stages of development. From the data presented above it is apparent that the placenta of the young hamster embryo is permeable to many of these teratogenic heavy metals. While this is suggestive that the teratogenic effect of heavy metals may be directly on the embryonic systems, we cannot yet exclude the possibilities that the teratogenic mechanism may be focused on the maternal organism or directly on the placenta.

With the demonstration that these heavy metals have specific teratogenic effects, new possibilities exist for the study of the ontogeny of specific malformations. Radioautography, EM histochemistry, and electron probe analysis are readily available techniques which offer opportunities for study unique to these "molecular teratogens." Hopefully, these molecular teratogens will offer new insight into the problems of developmental malformations.

Acknowledgment: This work was supported by USPHS grant ES-00697.

References

Brinster, R.L., Cross, P.C.: Effect of copper on the pre-implantation mouse embryo. Nature (Lond.) 238, 298-399 (1972).

Carpenter, S.J., Ferm, V.H.: Embryogenesis of sacral-tail malformations in lead-treated hamsters. Anat. Rec. 178, 323 (1974).

Carpenter, S.J., Ferm, V.H., Gale, T.F.: Placental permeability of inorganic lead in the golden hamster: Radioautographic evidence. Experientia (Basel) 29, 311-313 (1973).

Chang, C.C., Tatum, H.J., Kincl, F.A.: The effect of intrauterine copper and other metals on implantation in rats and hamsters. Fertil. and Steril. 21, 274-278 (1970).

Dutt, B., Mills, C.F.: Reproductive failure in rats due to copper deficiency. J. comp. Path. 70, 120-125 (1960).

Ferm, V.H., Carpenter, S.J.: Developmental malformations resulting from the administration of lead salts. Exp. molec. Path. 7, 208-213 (1967).

Ferm, V.H.: Developmental malformations induced by cadmium: A study of timed injections during embryogenesis. Biol. Neonate 19, 101-107 (1971).

Ferm, V.H., Carpenter, S.J.: The relationship of cadmium and zinc in experimental mammalian teratogenesis. Lab. Invest. 18, 429-432 (1968).

Ferm, V.H., Carpenter, S.J.: Teratogenic and embryopathic effects of indium, gallium, and germanium. Toxicol. appl. Pharmacol. 16, 166-170 (1970).

Ferm, V.H., Ferm, D.W.: The specificity of the teratogenic effect of lead. Life Sci. 10, 35-39 (1971).

Ferm, V.H., Hanlon, D.P.: Toxicity of copper salts in hamster embryonic development. Biol. Reprod. 11, 97-101 (1974a).

Ferm, V.H., Hanlon, D.P.: Placental transfer of zinc in early embryogenesis. J. Reprod. and Fertil. 39, 49-52 (1974b).

Ferm, V.H., Hanlon, D.P., Urban, J.: The permeability of the hamster placenta to radioactive cadmium. J. Embryol. exp. Morph. 22, 107-113 (1970).

Ferm, V.H., Saxon, A., Smith, B.W.: The teratogenic profile of sodium arsenate in the golden hamster. Arch. env. Hlth 22, 557-560 (1971).

Grüneberg, H.: The Pathology of Development. Oxford: Blackwell Scientific Publications 1963.

Holmberg, R.E., Ferm, V.H.: Inter-relationships of selenium, cadmium and arsenic in mammalian teratogenesis. Arch. env. Hlth 18, 873-877 (1969).

Hurley, L., Swenerton, H.: Congenital malformations resulting from zinc deficiency in rats. Proc. Soc. exp. Biol. (N.Y.) 123, 692-696 (1966).

O'Dell, B.L.: Trace elements in embryonic development. Fed. Proc. 27, 199-204 (1968).

Zipper, J., Medel, M., Prager, R.: Suppression of fertility by intrauterine copper and zinc in rabbits. Amer. J. Obstet. Gynec. 22, 107-113 (1970).

V. Cytogenetics

Experimental Heteroploidy in Mammals

O. BOMSEL-HELMREICH

I. Introduction

The chromosomes transmit the genetic information from one generation to the next; any alteration in their number or organization is bound to be of consequence. This is especially so in the performance of the reproductive system: chromosome aberrations are connected with sterility or reduced fertility, embryonic and fetal mortality, congenital malformations, and perinatal death.

Many workers have contributed in the last 20 years to the exploration of the field of spontaneous or induced heteroploidy (see *Fechheimer*, 1972, for recent review). Experimental induction was feasible in three ways.

The first is the induction of chromosomal aberrations of any type by various exogenous factors. Ionizing radiations are known to induce chromosome breaks. Chemicals have also been extensively used to induce various numerical aberrations. But their action is not limited only to DNA and chromosomes but to other cellular constituents which also play a role in embryonic development. Therefore this type of experiment is better able to give information on the influence of chromosome anomalies on cells than on individuals. To get information on the development of embryos with an abnormal karyotype, two methods are conveniently employed: (1) the induction of physiological situations known to increase embryonic mortality and, therefore, also chromosomal aberrations, and (2) the induction of programmed aberrations of one or a complete set of chromosomes.

II. Induction by Preferential Physiological Situations

The experimental reproduction of these situations opens a valuable field; the most widely and recently explored is the aging of gametes.

The importance of such experiments is confirmed by the possibility that these situations may occur frequently in farm animals or human reproduction. The right timing of mating or insemination and the use of hormones to induce ovulation is important either to regulate reproduction or to increase egg production. Furthermore, in man, long or irregular cycles appearing in the very young or premenopausal women often relate to spontaneous abortion and chromosomal abnormalities. They appear in relation with a delay in ovulation (*Jongbloet*, 1971). As for delayed fertilization, it is bound to occur frequently in a species such as the human, where the possibility of coitus is not limited by an estrous behavior of the female.

1. The Aging of the Egg

The egg has a relatively short life span. When the oocyte, ready to be ovulated, remains inside the follicle, aging occurs after 24 h. Once the egg is released outside the follicle, aging takes only a few hours.

a) Overripeness of the Oocyte

If ovulation of a ripe oocyte is delayed for 24 h in the rat by an injection of sodium pentobarbital, chromosomal aberrations in the postimplantation embryos increase three times (almost 5%). Trisomics, monosomics, triploids, tetraploids, and chimeras are observed. The embryonic mortality increases as well (38% at 11 days of gestation) (*Butcher* and *Fugo*, 1967). In the rabbit, follicular aging of the oocyte occurs when ovulation is delayed for 60 h after the last injection of follicule stimulating PMSG (*Bomsel-Helmreich*, 1972). The number of chromosomal anomalies and embryonic mortality is much higher than in the rat because observation takes place earlier, at the blastocyst stage. Of the blastocysts, 66% are abnormal, 20% have an abnormal karyotype; anomalies are the same as in the rat.

Especially remarkable is the occurrence of monozygotic twins: two blastocysts with embryonic discs and blastocoeles inside the same zona pellucida (1.5%). They develop up to postimplantation stages in the rat as observed by *Butcher* and *Fugo* (1967). Very large polar bodies have been observed to occur in the rat which could determine monozygosity. However, the described twins more likely have their origin in the separation of the first two blastomeres. Another proof that anomalies caused by overripeness may materialize as late as the first division is the high number of chromosomal chimeras found in the rat as well as in the rabbit.

b) Underripeness of the Oocyte

In a parallel study, *Iwamatsu* and *Chang* (1971) have shown in mice that eggs can develop into young even if they are recovered in the follicles as soon as 3 h before normal ovulation, and then fertilized and transplanted. The block to polyspermy does not function at this moment, it develops later. Efficient blocking increases progressively up to the time of normal ovulation. Consequently, underripeness, i.e. ovulation before the normal time, may also be considered as a source of heteroploids.

Fujimoto et al. (1974) observed in the rabbit an increase of heteroploid blastocysts after a medium superovulation (10%); one third of the embryos were triploid. Superovulation induces chromosome anomalies probably by an accelerated maturation of an oocyte reserve and because of ovulation of underripe eggs.

c) Aging of the Ovulated Egg

The aging of eggs has been observed mostly immediately after fertilization.Digyny and dispermy are the most often described abnormalities, possibly because they are the easiest to observe. Both have been extensively studied in different species. Their percentage may be quite high: 34% of eggs are digynic in the hamster (*Chang* and *Fernandez- Cano*, 1958) and 21% in the sow; 10% suffer dispermy in the sow, (*Thibault*, 1959). The observation of later stages of development shows not only that these eggs are able to develop but also that lesser mitotic anomalies occur in a much higher proportion. *Austin* (1967) obtained 25% triploids and 65% trisomic and monosomic blastocysts in the rabbit; *Shaver* and *Carr* (1967) and *Vickers* (1969) obtained them in the mouse, and *McFeely* (1967) in the sow. Mixoploids appear as well. In the sow at preimplantation stage, besides a high embryonic mortality, 25% of the embryos are morphologically abnormal and retarded. A certain number are triploid but the large majority are diploid (*Bomsel-Helmreich*, 1967). The relative part played by changing environmental factors (age) with delayed mating and the aging of the egg per se can be studied efficiently using in vitro techniques. In the rabbit *Thibault* (1967) aged eggs in vitro and fertilized them with aged sperm. Digyny was observed in more than half of the eggs and dispermy in 20%. This shows a direct effect of aging on the games. *Fraser* and *Dandekar* (1973), also in the rabbit, aged the eggs in vitro but fertilized them with fresh sperm and cultured them for two days. They observed no reduction in the fertilization rate and no increase in triploid embryos; nevertheless there was a much reduced vitality after the first cleavage. The authors assume that the effects of delayed mating were mainly due to tubal factors. In the rabbit *Bomsel-Helmreich* and *Szöllösi* (1974) transferred freshly ovulated eggs into aged tubes where they were fertilized. They demonstrated that aging of the tubes has some effective influence on fertilization rate and embryonic mortality but to a much lesser extent than the aging of the egg per se.

2. Aging of Sperm

It has been known for more than 10 years that the capacity of sperm to fertilize and to support normal embryogenesis decreases with time between the arrival of sperm in the female genital tract and fertilization.

Tesh (1966) demonstrated both pre- and postimplantation losses in the rabbit; with increasing age of the spermatozoa the losses appeared at an earlier stage. *Maurer* et al. (1969) reported reduced cleavage in cultured embryos taken from rabbits inseminated 20 h before ovulation. *Thibault* (1967) described digyny in 40% of freshly ovulated eggs fertilized with aged sperm. *Martin* and *Shaver* (1972) observed the chromosomal constitution of rabbit blastocysts when sperm was aged in utero for 14-28 h. Almost 10% had abnormal karyotypes: mixoploids (2n/4n), mosaics (2n/2n-1), or chimeras (XX/XY). It is interesting that digyny, which is an anomaly of the egg and may be considered as an incomplete activation, is induced in a young egg by aged sperm. Nondisjunction as late as the first cleavage is also observed.

Another type of aging occurs when vaginal sperm is used for in vitro fertilization; these spermatozoa seem to penetrate more slowly than uterine sperm. Some fertilization result in triploid embryos (*Dandekar* and *Fraser*, 1974).

Dispermy appears in the rabbit when the interval between insemination and ovulation is abnormally long (*Harper*, 1970). This may be related to a higher number of sperm present at the site of fertilization. Similarly, postejaculatory storage of mammalian spermatozoa before their use in artificial insemination leads to a reduction of fertility and to an increase in embryonic mortality in the bull (review by *Salisbury* and *Hart*, 1970) and in the rabbit (*Stranzinger*, 1970; *Kofoed-Johnsen* et al., 1968). Storage possibly reduces fertility in such a way, that it leads to delayed fertilization, and that chromosome aberrations occur in the waiting ova.

III. Induction of Definite Chromosomal Aberrations

This section is concerned only with the induced variations of a complete set of chromosomes. The induction of parthenogenesis, triploidy, and tetraploidy by exogenous factors have been studied.

1. Parthenogenesis

Parthenogenic development may appear after exposure to various stimuli, such as hyper- and hypotonic solutions, cooling or heating (for review see *Beatty*, 1957, 1967). Many experiments were made on the rabbit, as in this species suppression of polar body is easily obtained by moderate cooling. *Chalmel* (1962) even obtained fertilization (though abnormal) of the parthenogenically activated egg.

In the mouse, *Braden* and *Austin* (1954) employed heat shock as the activating agent but did not continue observation after 70 h. Recently, *Graham* (1970) obtained development of mouse eggs up to the blastocyst stage by cultivating cumulus-free eggs in vitro. *Tarkowski* et al. (1970) obtained development beyond the stage of implantation as a result of activation in vivo with an electric current. *Komar* (1973) activated eggs by heat in vitro: fewer eggs developed to morula and blastocyst stage (15%) compared to the two preceding ways of induction (50%).

Graham (1970), *Tarkowski* et al. (1970), and *Witkowska* (1963) obtained mostly haploid parthenogenons, *Komar* (1973) obtained some diploids but her low percentage of success may be explained by the induction of many hypohaploids which seem to be unable to cleave. Whatever the means of production, parthenogenetic eggs developed more slowly than controls.

2. Triploidy

To induce triploidy, three experimental means exist: fertilization by two spermatozoa (dispermy), fertilization after inhibition of the expulsion of the second polar body (digyny), and fertilization by a diploid sperm.

a) Dispermy and Polyspermy

Penetration of the ooplasm by more than one spermatozoon has been described in many species including the rat, mouse, hamster, guinea pig, rabbit, and pig (for reviews see *Austin* and *Bishop*, 1957; *Piko*, 1961). *Austin* and *Braden* (1954) used heat shock and showed that exposure of the oviducts of estrous rats to a temperature of 45°C just before the expected time of fertilization increases the proportion of polyspermic eggs. *Gwatkin* and *Williams* (1974) report that when the hamster eggs are exposed to heat, polyspermy increases to 17% once the temperature has reached 40°C. Cortical granule reaction is inhibited and the penetration of more than one spermatozoon possible.

The domestic pig seems especially favorable for these studies, as polyspermy has been found to reach a striking incidence under a variety of experimental conditions. By delayed mating or delayed insemination (*Pitkjanen*, 1955; *Hancock*, 1959; *Thibault*, 1959) an alteration of the egg membrane seems to occur by which the aged egg is unable to perform a block to polyspermy. Polyspermy occurs also after injecting progesterone systemically (*Day* and *Polge*, 1968) or locally (*Hunter*, 1972) shortly before ovulation, or after inducing ovulation during the luteal phase of the estrous cycle (*Hunter*, 1967). The relaxation of the isthmus and utero-tubal junction occurs after local injection of progesterone. Polyspermy is then thought to originate principally from the increased number of spermatozoa reaching the site of fertilization and an almost simultaneous penetration of the egg membrane by two or more spermatozoa before completion of the zona reaction. The direct effect of progesterone on the membrane of the gametes, however, cannot be excluded.

In the pig, surgical resection of the isthmus is followed by the same high percentage of polyspermic egg (34%). This is true also to a much lesser extent in the rabbit (*Hunter* and *Leglise*, 1971). The direct exposition of excessive number of spermatozoa at the site of fertilization induces polyspermy as well (*Hunter*, 1973).

In both, dispermic and trispermic eggs, a single male pronucleus is uniting with the female pronucleus. The accessory male element(s) remains at the opposite hemisphere of the egg. *Gulyas* (1974) observed by electron microscopy in the rabbit that the supernumerary sperm form pronuclei; but they do not associate with the female pronucleus as they do in eggs observed after delayed fertilization by *Thibault* (1959). *Gulyas* concludes that, although the fully expanded acessory male pronucleus is similar to the pronuclei of monospermic fertilization, the temporal events of nuclear envelope formation are disturbed. This may leave doubt about the possibility of induction of polyploid embryos: until now the descendants of these polyspermic zygotes have not been studied.

If the simultaneous presence of a high number of sperm present at the site of fertilization is sufficient to induce polyspermy, in vitro fertilization could prove to be an ideal model.

This is not true for rabbit eggs. Only monospermic fertilization is observed even though the number of sperm present in the in vitro milieu largely exceeds the number present in vivo. Rabbit eggs are more prone to react with digyny. Little dispermy was observed by *Motlik* and *Fulka* (1974) with pig eggs and in vitro fertilization. This is surprising since pig eggs react easily to dispermy in vivo as described above.

In the hamster, polyspermy shows a low incidence (1.6%) after normal mating; after in vitro fertilization this incidence may be as high as 100% (*Yanagimachi* and *Chang*, 1964). In the majority there are more than two spermatozoa present, and it is not probable that these polyspermic eggs would be able to develop further. *Barros* et al. (1972) assume that these simultaneously present sperm heads are normal but at different stages of pronuclear transformation; but they never describe syngamy as in polyspermic fertilization of aged eggs.

When rat eggs are fertilized in vitro with epididymal sperm, 13% are polyspermic and none dispermic (*Niwa* and *Chang*, 1974). The authors use superovulated immature donors to obtain the eggs so that polyspermy may be related to a difference in the reaction of the zona pellucida in eggs of prepuberal ovaries (*Niwa* and *Chang*, 1973) as no polyspermy is observed otherwise.

b) Digyny

Triploid embryos have been produced at will by using colchicine or a related compound that suppresses cytokinesis: *Edwards* (1958, 1961) in the mouse, *Piko* and *Bomsel-Helmreich*

(1960) in the rat. They injected the compound into the mother during ovulation or fertilization. Heavy embryonic mortality and numerous abnormal retarded embryos were found mostly by direct damage of the alcaloid. A less damaging way is to use colcemid during a short and precisely defined time, i.e. at expulsion of the second polar body, if in vitro fertilization in the rabbit is used (*Bomsel-Helmreich* and *Thibault*, 1962; *Bomsel-Helmreich*, 1967). The success is almost 100% and when the eggs are transplanted into a foster mother the digynic embryos can be studied extensively.

c) Development of Triploid Embryos

This technique was used in a study concerning more than 1500 eggs treated in vitro and transferred into foster mothers (*Bomsel-Helmreich*, 1967, 1970). Descendants were obtained at different stages of gestation. Triploid blastocysts are smaller than diploid blastocysts of the same age; if too small at 160 h postcoitum, they are unable to implant. This is true equally for any abnormal blastocyst. Once this crisis has passed, lethality does not correspond to a definite developmental stage. In the rabbit, between the 10th and 13th day of gestation, mortality of triploids remains constant around 33% (embryos per egg transferred). On the 14th day, mortality raises to 63%, to 97% at 15 days, and 100% at 16 days.

Morphologic and anatomical data weight, and aspects of the embryos such as appearance of limb buds and general anatomy were collected to explain the reason of death. Nothing specific was observed; only a general delay in development appears clearly after 11 days, which increases to a difference in development of 18 h maximum between 3n and 2n embryos (Fig. 1).

The triploid heart seems slightly less developed than that of the diploid. But there is no clear-cut malformation that could explain death. The triploids do not stop their normal embryogenesis, but follow a normal course of development similar to that of diploids. Only a delay longer than 18 h seems to be incompatible with embryonic survival.

Is the mother-fetus relation normal in the heteroploid embryo? Examination of placenta of triploid rabbit embryos shows that their deterioration precedes the deterioration of the embryo by a short period of time. The placenta does not seem to be the cause of the delay of growth and final death. Neither is there a specific morphology typical of a special karyotype as in human (*Boué* and *Philippe*, 1969).

The possibility of deficiency in the endocrinological fetomaternal relationship was explored by different ways. Triploid and diploid eggs were transplanted together in the same foster mother. The growth of diploid embryos in the same mother was a definite stimulus for the growth of the triploids. When only 3n embryos are present, between 10 and 16 days, the proportion of triploid embryos to triploid eggs transplanted is less favorable (22%) than when diploid and triploid embryos are present (33%). The survival rate (living embryos/total embryos) is 3% for days 15-16 in the first case, and increases to 14% in the second case. Supplementation in progesterone improves embryonic survival of 3n embryos in the same proportion, the survival rate being 22% at 15-16 days. The same effect may be obtained again by a superovulation of triploid eggs alone (transfer of 15-38 eggs per doe). Despite the heavy but erratic embryonic mortality of triploid embryos, a sufficient number develops to stimu-

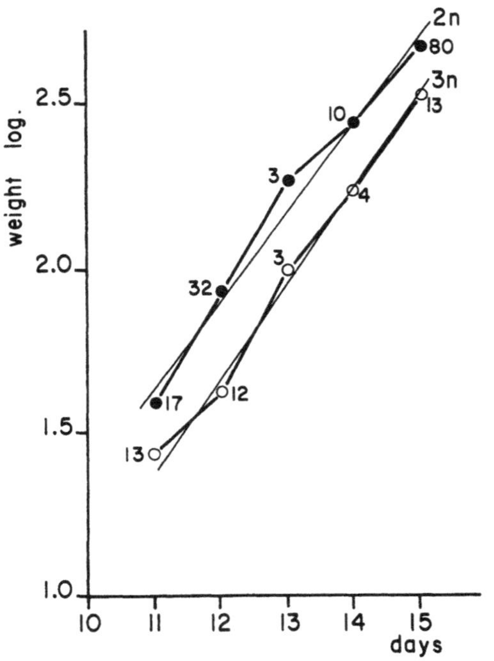

Fig. 1. Compared weight of 3n and 2n rabbit embryos from 11 to 15 days of gestation. Triploids show a 18 h delay of growth

late the foster mother and they may be carried almost to term. Living triploid embryos were obtained in the last quarter of gestation. These three methods: simultaneous transplantation of 3n and 2n eggs, superovulation of 3n eggs, and progesterone supplementation seems to overcome an insufficient stimulation given by triploid embryos to their mother.

The longer intra-uterine survival of human triploid embryos threatening abortion has also been obtained by hormonal supplementation (*Papiernik-Berkhauer*, 1968). But neither in rabbit nor human is this insufficient stimulation the principal cause of the developmental delay, moreover, the hormonal supplement cannot prevent the final mortality of the embryo. The main cause seems to be related to cell metabolism.

In human, spontaneous triploidy is frequent: 20% of spontaneous abortions with chromosomal anomalies are 3n; more than 300 cases of triploid individuals, embryos, or children have been described. Before 20 weeks of gestation, 275 triploid aborted, 22 survived 27 weeks, and 18 survived 28 weeks of gestation; none survived birth more than a few hours or days (*Niebuhr*, 1974). Information obtained from human triploids is parallel to that obtained experimentally in the rabbit. In spontaneous abortion of the first trimester triploids live longest, compared to any other chromosomally abnormal embryo. There is no specific mechanism that explains why the majority aborts early but some are able to survive till birth.

The genetic means to obtain triploids was used by *Wroblewska* (1971). She crossed inbred female animals with males of a different strain. They showed all a specific disturbance of embryogenesis, mainly an inhibition of the embryonic part of the egg cylinder to form; their origin was probably digyny.

d) Triploidy by Diploid Sperm

Another approach to induce triploidy in excellent in vivo conditions is suggested by *Beatty* and *Fechheimer* (1972); in certain individuals, rabbits, humans and other species, diploid sperm appear quite often. They could be separated from haploid sperm by centrifugation and after artificial insemination they would induce triploid zygotes. *Fechheimer* and *Beatty* (1974) demonstrated that in the rabbit the insemination with semen containing 3% of diploid sperm does not increase the number of triploid blastocysts. This suggests that triploidy does not often find its origin in nonreduced male gametes.

3. Tetraploidy

In contrast to triploidy or haploidy, a balanced polyploidy seems to be compatible with complete fetal development in some mammals. Recently, *Snow* (1973) obtained tetraploid mouse embryos by culturing two-cell eggs in cytochalasine, which prevents cytokinesis without influencing nuclear division. The proportion of treated embryos that develop normally to blastocyst stage is important; they do implant but most have a very limited development; some show normal development and may even develop to term.

In similar experiments, *Tarkowski* and *Witkowska* (personal communication) have found that in tetraploid egg cylinders the formation of fetal membranes proceeds normally but the embryo itself does not develop. This is similar to what is described for triploid mouse embryos by *Wroblewska* (1971).

Tetraploids obtained by cell-fusion (*Graham*, 1971) develop very little. *Tarkowski* (personal communication) also uses a fusion technique to obtain tetraploid blastocysts. Most of the tetraploid embryos are abnormal; but as the size of tetraploid cells is larger than diploids, it is likely that the cause of embryonic malformations is in the reduced number of cells of the blastocysts.

IV. Mechanisms of Occurrence of Chromosomal Aberrations

Despite the numerous ways of induction, not much is known yet about the causal relation of chromosomal anomalies with abnormal development and embryo mortality.

Exogenous agents such as x rays or drugs generally induce chromosome aberrations at random. When the "physiological pathways of induction" are used almost all kinds of abnormalities take place such as digyny as well as dispermy, monosomy, trisomy, and chimeras. The precise time of appearance of the abnormality is not easy to define. Trisomies or monosomies may occur at the first or second meiotic division of the gametes. Triploidy may be induced by non-expulsion of the first or second polar body, or by dispermy. Mosaics arise at early cleavage but not necessarily at the first. There exists also

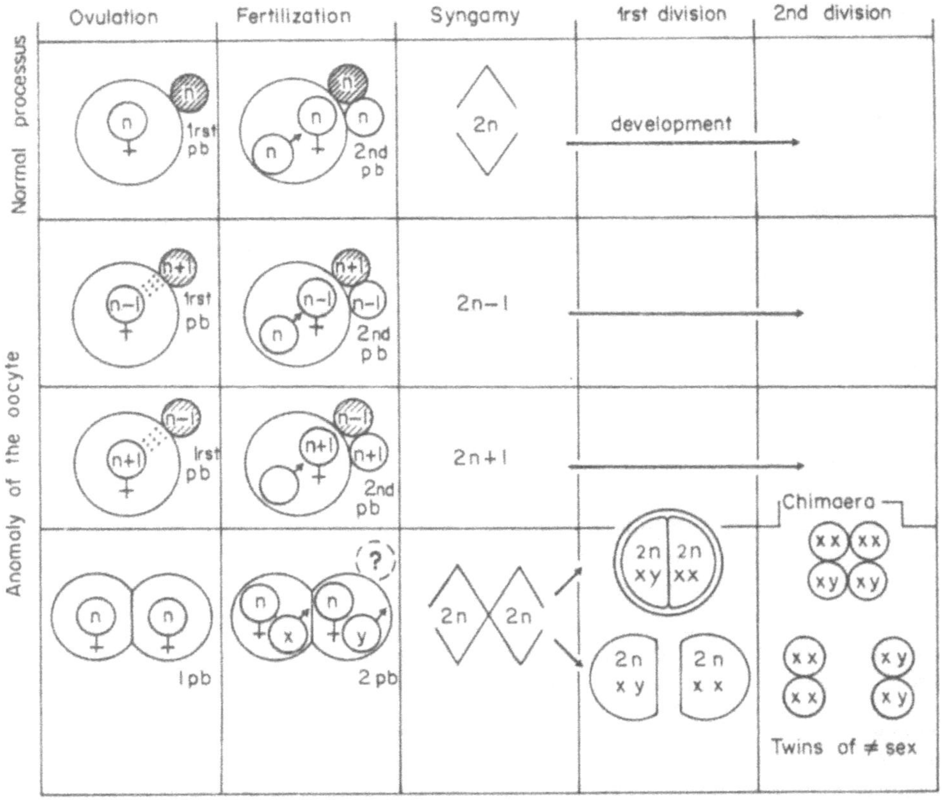

Fig. 2. Comparison of normal process of ovulation, fertilization, and first divisions and possible anomalies of ovulation; aneuploidy after non-disjunction at expulsion of 1st polar-body, chimeras or twins of possibly different sex after equation division instead of expulsion of 1st polar-body

the notion of related anomalies, that is to say, the appearance of different types of aberrations occurring subsequently: for instance successive nondisjunction, triploidy, and chimerism resulting in an embryo 3n-2/2n-1. This is explicable from a genetic point of view as well as from a physiologic one. The various types of chromosomal aberrations appearing in the oocyte, the ovulated, or the cleaving egg, and their causes are illustrated in Figures 2–5. The identification of the origin of individual chromosomes by the banding technique should be a great help in these problems.

The induction of a specific anomaly by genetic factors or of polyploidy by in vitro experiments gives a much better control over these variations. Nevertheless, it cannot suppress genetic regulations that are still not understood, such as the preferential survival of 3nXXY embryos to 3nXXX even when both are of digynic origin. The 3nXYY embryos seem to be especially prone to die.

It is equally possible that a different time of induction produces the same abnormality but does not allow an equivalent development. Such may be the case for instance in dispermy induced in normal eggs of the pig where the three pronuclei do not fuse; whereas dispermy

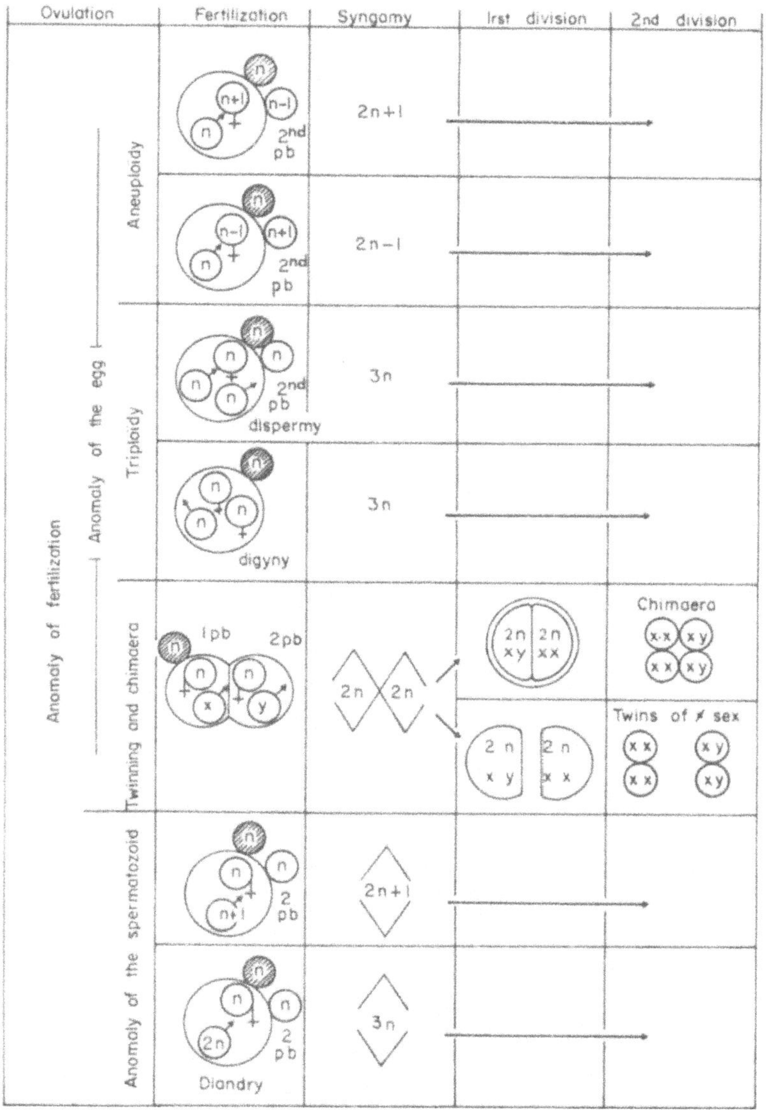

Fig. 3. Possible anomalies of fertilization: aneuploidy after non-disjunction at expulsion of the 2nd polar body–triploidy after digyny or dispermy, twins or chimeras after equational division instead of expulsion of the 2nd polar body–aneuploidy or triploidy after fertilization by an abnormal sperm

in an aging egg leads to a syngamy of the three pronuclei and therefore to a triploid embryo. Further, the way of induction does not necessarily induce only chromosomal aberrations. It may also lead to, as yet, insufficiently explored anomalies of the whole cell. This is easily accepted for different exogenous agents, but it occurs also in the case of physiological aging of the cytoplasm. This process determines chromosome anomalies but also the high mortality of diploid embryos (*Bomsel-Helmreich*, 1967).

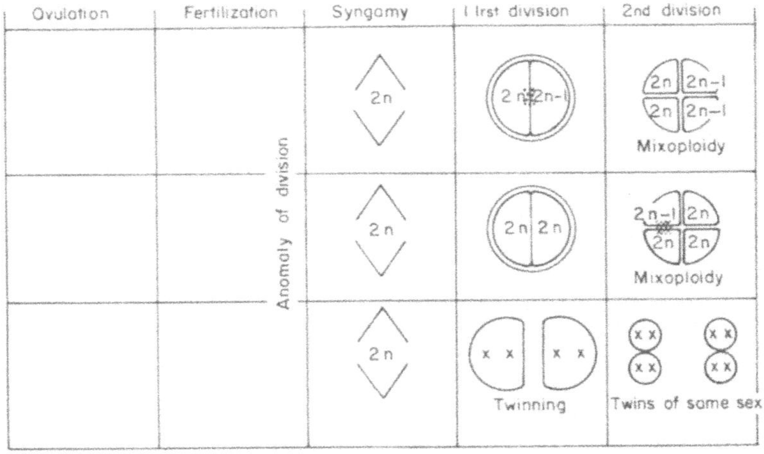

Fig. 4. Possible anomalies of division: mixoploidy after non-disjunction at 1st or 2nd division – twins, necessarily of same sex

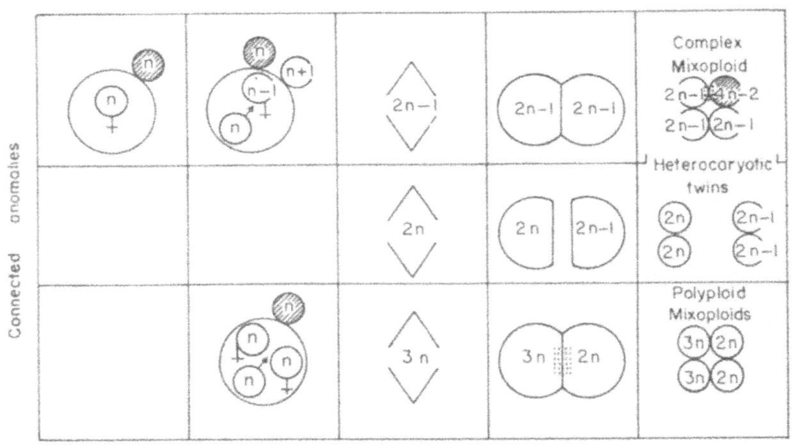

Fig. 5. Connected anomalies: complex mixoploids after non-disjunction at 2nd polar body and non-disjunction at 2nd division, – heterocaryotic twins after non-disjunction and twinning at first division · polyploid mixoploids after non-expulsion of 2nd polar body and loss of set of chromosomes at first division

V. Causes of the Mortality of the Triploid Embryo and Characteristics of the Triploid Cell

A most puzzling topic remains – the almost complete lack of information of the causes of embryonic mortality of chromosomally abnormal embryos. In some cases the embryo shows a gross developmental malformation, but in most cases it is only delayed in growth.

Very few causal studies have been undertaken yet. The triploid cell, because of its balanced chromosome anomaly and well-defined origin, is of special interest in this respect.

Some data concerning different morphologic and physiologic criteria of cell activity have been obtained on experimental material. Cultured in vitro, 3n and 2n fibroblasts are obtained from 10-12 day rabbit embryos, triploidy having been induced by colcemid after in vitro fertilization (*Jerôme* et al., 1970; *Bomsel-Helmreich*, 1972; *Jerôme* et al., 1976). The results will be compared with the information available from human triploid spontaneous abortions. It should be kept in mind that we compare cellular mechanisms in triploids obtained experimentally from normal (young) eggs with spontaneous triploid originating mostly from defective (aged) eggs.

1. Length of Mitotic Cycle

Mittwoch proposed in 1967 that trisomic cells had a longer S or G2 phase than diploid controls. She measured the amount of DNA in Feulgen-stained cells by microphotometry. In tri-21 cultures, a larger fraction of the cell population was in S or G2 phase than in 2n cells. A similar technique was used to demonstrate that triploid cells had a longer cycle time than diploids (*Mittwoch* and *Delhanty*, 1972). *Paton* et al.(1974) secured these findings by using labeled mitoses in culture. In trisomic human fibroblasts, the G2 phase is 3-4 h longer than in normal controls.

Cure et al. (1973) compared doubling time of 3n and 2n fibroblasts in culture obtained from spontaneous abortions. The average doubling time for 3n cells was 32 h, that for monosomic and trisomic being 37 and 36 h, respectively. The average for diploid cells was 35 h but doubling time of fibroblasts obtained from therapeutic abortions was definitely shorter and corresponded to doubling times of normal diploid fibroblasts (17-25 h).

The life span of these different types of cells is also shorter in the same proportion: 15-24 generations for aneuploids, 28 generations for triploids, against 40-60 in normal diploid cells. In rabbit triploid cell cultures this shortening of life span has also been observed (*Jerôme* et al., 1976).

Schneider and *Epstein* (1972) previously reported that skin fibroblasts derived from patients with trisomy 21 also had a significantly decreased number of cell population doublings.

These observations may also be important for an understanding of the gradual disappearance of aneuploid or triploid cells in mosaic tissues. *Mittwoch* and *Delhanty* (1972) studied a mixture of 2n/3n fibroblasts in culture. The proportion of triploid cells fell in time and no mitoses were observed. This phenomenon is likely to occur also in vivo. Although these fibroblast cultures originated from a skin biopsy containing a mixture of 3n/2n cells, a blood culture from the same patient contained only diploid cells. In three other 2n/3n patients, the 3n component was also confined to cells derived from skin and fascia lata; dividing cells from lymphocytes and bone marrow were found to be exclusively diploid (*Böök* and *Santesson*, 1961; *Ferrier* et al., 1964; *Schmid* and *Vischer*, 1967). *Ford* (1964) suggested that the proliferous ability of normal and aneuploid cells may differ between rapidly dividing tissues such as bone marrow and slowly dividing elements such as dermis. The investigations suggest that slow cell growth may confer an overall handicap to the triploid condition because of this lengthening of any or all phases of mitotic cycle.The triploid cell can participate in normal embryonic development except for certain critical stages, where a rapid succession of divisions becomes necessary. This could explain not only general retardation in growth, but also the heart anomalies in human triploids who die near birth. In contrast, the 3n embryo at an early stage is generally delayed but shows a relatively harmonious development.

The same concept applies to any trisomic individual. It could explain certain similarities of malformations such as colomboma, heart deficiencies, anomalies of gonads, variations of dermatoglyphs, facial deformation, and mental deficiency, together with a general delay of growth between individuals with different types of trisomy or with triploidy.

2. Protein Content and Enzymatic Activities of Triploid Cells

The increase of genetic material in any heteroploid, but especially in triploids should have an effect on cell metabolism. This led to a study of some enzymatic activities and other

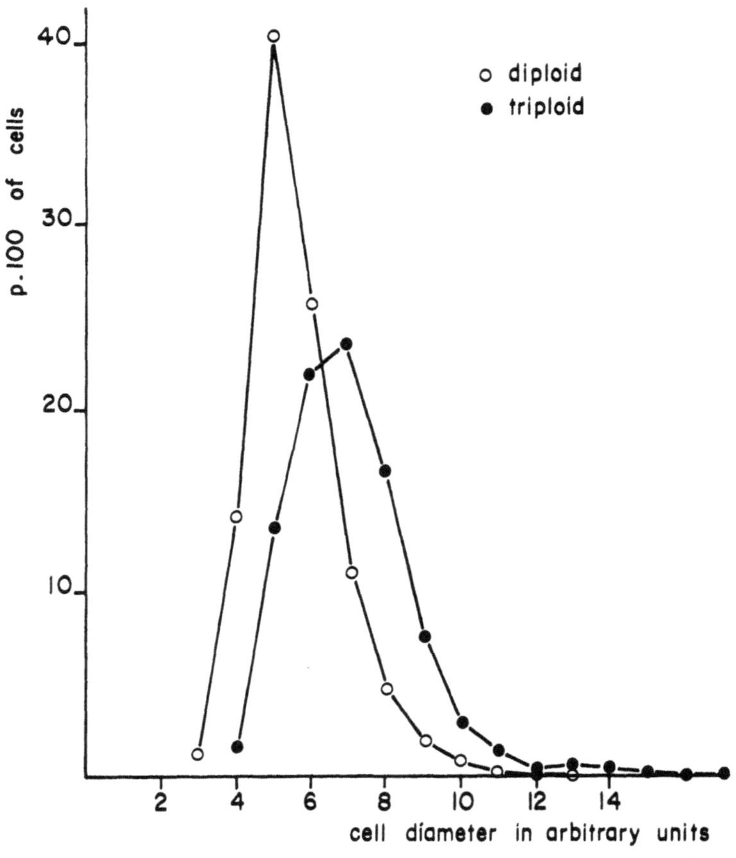

Fig. 6. Comparison of cell diameters of triploid and diploid fibroblasts of rabbit embryos in culture

related parameters on 3n and 2n rabbit fibroblasts in vitro, obtained from 10-12 day rabbit embryos (*Jerôme* et al., 1970; *Bomsel-Helmreich*, 1970; *Jerôme* et al., 1976).

As expected, the quantity of DNA is significantly higher in triploid cells, in a relation 3n/2n 1·63. Values are more dispersed in triploids. The diamters of more than 8000 cells were measured. The difference of diameter is highly significant; triploid cells are larger, but the range is also large, and the smallest 3n cells are almost of the same size as the smallest 2n cells.Some triploid embryos have all their cells of a diploid size (Fig. 6) and these embryos show an improved viability; in vivo, they are larger embryos than normal triploid embryos, and almost as large as diploids. This improved viability persists during culture. These strains prosper as well as diploids, at least definitely better than other triploids. Total protein content was determined by Lowry's method on a large number of cells (100,000 or 50,000). 66 determinations were made on cells from 17 triploid embryos and 78 determinations from 26 diploid embryos. Despite a large dispersion of values in triploids, no difference was observed in the protein content per cell between 3n and 2n in two different series of determination (Fig. 7). These results were independent of the number of passages in

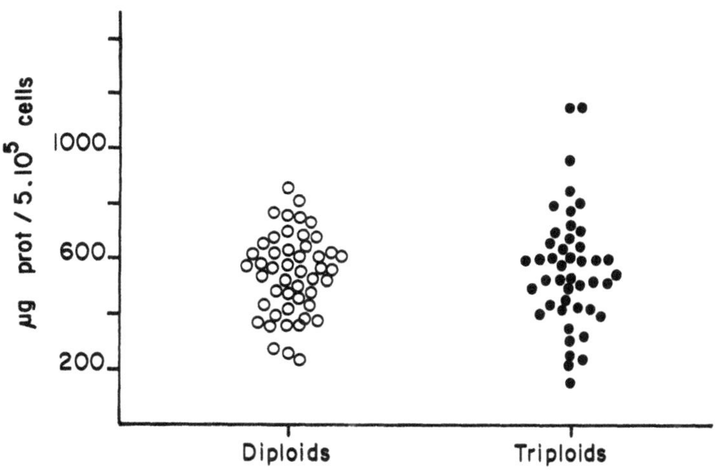

Fig. 7. Comparison of total protein content per cell in triploid and diploid fibroblasts of rabbit embryos in culture

culture. Nine enzymatic activities were determined. First an X-linked enzyme (Glucose 6 phosphodehydrogenase (G6PD), and four autosomal enzymes (phosphogluconate dehydrogenase (6PGD), phosphofructokinase (PFK), galacturidyltransferase (Gal-Trans), and acid phosphatase). In 77 assays of diploids and 109 of triploids the activity of 6PGD, PFK, and Ac.PH were not different between 3n and 2n cells; Gal-trans was lower in triploids (3n/2n 0.65). The X-linked G-6-PD was higher in triploids (3n/2n 1.86). The second series (539 determinations) was a comparison of three X-linked enzymes: G-6-PD, Hypoxanthinphosphoribosyl-transferase (HPRT), and phosphoglycerate kinase (PGK), and three autosomal enzymes: 6PGD, pyruvate kinase (PK), and adeninephosphoribosyl-transferase

(APRT). In this comparison no difference was observed in the activities of those enzymes determined by autosomes, but a highly significant difference in the X-linked enzymes was found.

Two conclusions can be drawn: there are differences between enzymatic activities when linked to autosomes: five enzymes tested showed no difference between 3n and 2n cells, but one (Gal-trans) had a lower activity in triploids. These anomalies of the enzymatic balance could be in part an explanation for the delay in growth or the progressive embryonic mortaility in triploids.

X-linked enzymes have almost twice the activity of those determined by genes. This increase suggests a relation with an abnormal inactivation of the 2X chromosomes which are always present in the triploid embryos because of their digynic origin. Observation of nuclear sex shows that all 3A XXY strains lack a Barr body; however, the 3A XXX possess one, and in very low percentage two. This would account for the higher activity of X-linked enzymes in triploids.

3. Nuclear Sex in Triploids

The just cited findings raise the question of the expression of nuclear sex in triploid cells. In human triploids (abortuses or children dying near birth) almost 50 3A XXX and 3A XXY have been examined for Barr bodies. A variable but significant number of 3A XXX cells are sex chromatin-positive; only a small number have a second Barr body usually in less than 5% of the cells. Similarly, half of the XXY have no sex chromatin, the others have one Barr body but in less than 3% of the cells. A majority of the XXX embryos has Barr bodies to a much higher extent (50 to 90% of cells). These inconsistent results show that the number of Barr bodies in a cell depends not only on the number of X chromosomes but also on the ploidy of the cell. The relationship between the maximum number of Barr bodies (B), X chromosomes (X) and haploid sets (P) is given in the following equation: $B = X - P/2$ (*Harnden*, 1961). In the case of 3 AXXX cells the results would be 1.5 Barr bodies and 0.5 for 3A XXY.

The unstable chromosome effect on Barr bodies observed in human triploids compared to the stable inactivation observed in the rabbit is puzzling. Two main differences exist between the two series: the rabbit cells are all from digynic origin, whereas the human triploids probably are either of digynic of dispermic origin. The origin of the extra haploid set may ultimately be connected to its allocyclicity. Thus, if two chromosome sets are of maternal origin, (digyny) only one X chromosome would be inactivated. Conversely, if two chromosomes are of paternal origin (dispermy) two X chromosomes would be inactivated for 3A XXX, and one for 3 AXXY. The hypothesis advanced by *Brown* and *Chandra* (1973) gives no explanation why triploid individuals with two Barr bodies have not been observed, even though it is probable that the majority of human triploid embryos are of digynic origin. *Nikawa* and *Kajii* (1974) determined by the fluorescence and banding techniques the dispermic origin of a triploid 3A XXY human embryo and despite this origin, it was sex chromatin negative.

The second difference between the rabbit and human triploid cells is their origin. In the human the original egg cell could have a decreased vitality, as it develops probably from an aged gamete. Triploidy in rabbits is artificially obtained from a fresh egg. This could explain the regularity of inactivation in the rabbit cells and the erratic inactivation in human cells. Other cellular anomalies are observed even in diploid cells obtained from human abortions. Thus, the doubling time of embryonic diploid cells is longer when obtained from spontaneous abortions (over 30 h) than when obtained from induced abortions (17-25 h).

Human triploid infants with XXY sex chromosomes tend to be intersexual; the same sex chromosomes in a diploid karyotype led to the male phenotype of Klinefelter's syndrome (2AXXY). This feature may be the result of a slowing of growth in triploids with the assumption that male development requires faster growth than female development (*Mittwoch* et al., 1969). It may also be related to the failure of inactivation of the second X chromosome as demonstrated in the triploid rabbit.

In conclusion of this first approach to an explanation of embryonic mortality, triploid cells have a larger size and more DNA than diploid cells, but they have the same quantitiy of protein. They undergo division at a different rhythm than diploid cells but to a lesser extent than the less balanced trisomic or monosomic cells. The enzymatic activities are definitely disturbed when X-linked, which is related to an abnormal inactivation of X chromosomes, but this also exists in autosomal linked enzymes. These effects seem to be apparent also in the even better-balanced tetraploid embryos such as those obtained by *Snow* (1973) which, despite larger cell size, survive sometimes to birth.

VI. Conclusion and Summary

Experimental heteroploidy brings new light to very different and important aspects of the physiology of ovulation and fertilization.

The induction by "preferential physiological situations" shows that there are multiple ways to induce the same chromosomal anomaly and like ways to induce multiple anomalies. Knowledge of these phenomena is important for the prevention of spontaneous and undesired disturbances of embryonic growth in man and other mammals. Many aspects need to be explored that would yield information on the physiologic process itself.

The induction of programmed aberrations is a precise and elegant tool. It is not only a way to obtain embryos or individuals with only one chromosome anomaly, but free from the effects of factors connected with the spontaneous appearance of heteroploids, such as aging. It is also a better way to explore the effect of known numerical aberrations on the cell metabolism. This last point is also the newest consideration. Why do heteroploid embryos die? How do they die? The exploration of gametic and metabolic causes of this mortality is certain to bring essential and new information about cell behavior under an abnormal chromosome constitution and also about the effects of alteration of cell activities, such as cycle length on critical moments of embryogenesis, where the limits of tolerance of embryonic growth are very strict. It opens a new field of experimental embryology.

References

Austin, C.R.: Chromosome deterioration in aging eggs of the rabbit. Nature (Lond.) 213, 1018-1019 (1967).

Austin, C.R., Bishop, M.W.H.: Fertilization in mammals. Biol. Rev. 32, 296-349 (1957).

Austin, C.R., Braden, A.W.H.: Induction and inhibition of the second polar body division in the rat egg and subsequent fertilization. Aust. J. biol. Sci. 7, 195-199 (1954).

Barros, C., Vliegenhart, A.M., Franklin, L.E.: Polyspermic fertilization of hamster eggs in vitro. J. Reprod. Fertil. 28, 117-120 (1972).

Beatty, R.A.: Parthenogenesis and polyploidy in mammalian development. London: Cambridge University Press 1957.

Beatty, R.A.: In: Parthenogenesis in Vertebrates in Fertilization (eds. *Metz, Monroy*), Vol. 1, p. 413. New York: Academic Press 1967.

Beatty, R.A., Fechheimer, N.S.: Diploid spermatozoa in rabbit semen and their experimental separation from haploid spermatozoa. Biol. Reprod. 7, 267-277 (1972).

Böök, J.A., Santesson, B.: Nuclear sex in triploid XXY human cells. Lancet 1961 I, 318.

Bomsel-Helmreich, O.: Triploïdie expérimentale chez quelques mammifères. Thèse Doctorat d'Etat, Paris, p. 112 (1967).

Bomsel-Helmreich, O.: Fate of heteroploid embryos. Advances in Biosciences. 6. Schering Symp. on Intrinsic and Extrinsic Factors in Early Mammalian Development (ed. *Raspe*) pp. 381-403. Oxford-Braunschweig: Pergamon Press, Vieweg 1970.

Bomsel-Helmreich, O.: Effects de l'ovulation retardée sur le développement des blastocystes de Lapine. VII. Int. Kongr. für tierische Fortpflanzung, München, Band II, 205-210 (1972).

Bomsel-Helmreich, O., Szöllösi, D.: The effect of environment on fertilization following delayed mating in the rabbit. Abstr. Soc. Study Fertil.(1974).

Bomsel-Helmreich, O., Thibault, C.: Fécondation in vitro en présence de colchicine et polyploïdie expérimentale chez le Lapin. Ann. Biol. anim. Bioch. Biophys. 2, 13-16 (1962).

Boué, J.G., Philippe, E.: Les caractères anatomo-pathologiques des placentas humains selon les anomalies chromosomiques. C.R.Acad. Sci. (Paris) 269 (D) 535-537 (1969).

Braden A.W.H., Austin, C.R.: Reactions of unfertilized mouse eggs to some experimental stimuli. Exp. Cell Res. 7, 277-280 (1954).

Brown, S.W., Chandra, H.S.: Inactivation system of the mammalian X chromosome. Proc. nat. Acad. Sci. (Wash.) 70, 195-199 (1973).

Butcher, R.L., Fugo, N.W.: Overripeness and the mammalian ova. II. Delayed ovulation and chromosome anomalies. Fertil. and Steril. 18, 297-306 (1967).

Chalmel, M.C.: Possibilité de fécondation des oeufs de lapine activés parthénogénétiquement. Ann. Biol. anim. Bioch. Biophys. 2, 279-297 (1962).

Chang, M.C., Fernandez-Cano, L.: Effects of delayed fertilization on the development of pronucleus and the segmentation of hamster ova. Anat. Rec. 132, 307-320 (1958).

*Cure, S., Boué, A., Boué, J.:*Consequence of chromosomal anomalies on cell multiplication. In: Les accidents chromosomiques de la Reproduction, pp. 95-109. Paris: Coll. INSERM 1973.

Dandekar, P.V., Fraser, L.R.: In vitro fertilization of rabbit eggs using vaginal sperm. Abstr. Soc. Study Fertil. (1974).

Day, B.N., Polge, C.: Effects of Progesterone on fertilization and egg transport in the pig. J. Reprod. Fertil. 17, 227-230 (1968).

Edwards, R.G.: Colchicine-induced heteroploidy in the mouse. I. The induction of triploidy by treatment of the gametes. J. exp. Zool. 137, 317-322 (1958).

Edwards, R.G.: Induced heteroploidy in mice: effect of deacethylmethyl colchicine on eggs at fertilization. Exp. Cell Res. 24, 615-618 (1961).

Fechheimer, N.S.: Causal basis of chromosome abnormalities. J. Reprod. Fertil. Suppl. 15, 79-98 (1972).

Fechheimer, N.S., Beatty, R.A.: Chromosomal abnormalities, and sex ratio in rabbit blastocysts. J. Reprod. Fertil. 37, 331-343 (1974).

Ferrier, P., Ferrier, S., Stalder, G., Buhler, E., Bamatter, F., Klein, D.: Congenital assymetry associated with diploid-triploid mosaicism and large satellites. Lancet 1964 I, 80-82.

Ford, C.E.: In: Cytogenetics of Cells in Culture (ed. *R.J.C. Harris*), p. 27. New York–London: Academic Press 1964.

Fraser, L.R., Dandekar, P.V.: The effect of aging on in vitro fertilization of rabbit eggs and subsequent embryonic development. J. exp. Zool. 184, 303-312 (1973).

Fujimoto, S., Pahlavan, N., Dukelow, W.R.: Chromosome abnormalities in rabbit preimplantation blastocysts induced by superovulation. J. Reprod. Fertil. 40, 177-181 (1974).

Graham, C.F.: Parthenogenetic mouse blastocysts. Nature (Lond.) 226, 165-166 (1970).

Graham, C.F.: Virus assisted fusion of embryonic cells. Karolinska Symp. Research Meth. Reproduct. Endocrinol. 3rd Symp. (ed. *E. Diszfalusy*), pp. 154-167 (1971).

Gulyas, B.J.: Electron microscopic observation on advanced stages of spontanous polyspermy in rabbit zygotes. Anat. Rec. 179, 285-296 (1974).

Gwatkin, R.B.L., Williams, D.T.: Heat sensitivity of the cortical granule protease from hamster eggs. J. Reprod. Fertil. 39, 153-155 (1974).

Hancock, J.L.: Polyspermy of pig ova. Anim. Prod. 1, 103-106 (1959).

Harnden, D.G.: Nuclear sex in triploid XXY human cells. Lancet 1961 II, 488.

Harper, M.J.K.: Factors influencing sperm penetration of rabbit eggs in vivo. J. exp. Zool. 173, 47-62 (1970).

Hunter, R.H.F.: Polyspermic fertilization in pigs during the luteal phase of the oestrus cycle. J. exp. Zool. 165, 451-460 (1967).

Hunter, R.H.F.: Local action of progesterone leading to polyspermic fertilization in pigs. J. Reprod. Fertil. 31, 433-444 (1972).

Hunter, R.H.F.: Polyspermic fertilization in pigs after tubal deposition of excessive numbers of spermatozoa. J. exp. Zool. 183, 57-64 (1973).

Hunter, R.H.F., Leglise, P.C.: Polyspermic fertilization following tubal surgery in pigs, with particular reference to the role of the istmus. J. Reprod. Fertil. 24, 233-246 (1971).

Iwamatsu, T., Chang, M.C.: Factors involved in the fertilization of mouse eggs in vitro. J. Reprod. Fertil. 26, 197-208 (1971).

Jerôme, H., Cathelineau, L., Bomsel-Helmreich, O.: Activités enzymatiques des cellules d'embryons de lapin diploïdes et triploïdes cultivés in vitro. VI Int. Congr. Anim. Reprod. Art, pp. 893-895. Paris: INSERM 1968.

Jerôme, H., Kamoun, P., Bomsel-Helmreich, O.: Enzymatic activities in diploid and triploid rabbit embryo fibroblasts in culture (in preparation 1976).

Jongbloet, P.H.A.: Mental and physical handicaps in connection with overripeness ovopathy. Acad. Proefschrift, Leiden, Holland (1971).

Kofoed-Johnsen, H.H., Pavlok, A., Fulka, J.: Effects of in vitro storage of spermatozoa on fertilizing efficiency and embryonic mortality in rabbits. Arsberetn Inst. Sterilitetsforsk K. Vet-og Landbohosk, 169 (1968).

Komar, A.: Parthenogenetic development of mouse eggs activated by heat-shock. J. Reprod. Fertil. 35, 433-443 (1973).

MacFeely, R.A.: Chromosome abnormalities in early embryos of the pig. J. Reprod. Fertil. 13, 579-585 (1967).

Martin, P.A., Shaver, E.L.: Sperm aging in utero and chromosomal anomalies in rabbit blastocyst. Develop. Biol. 28, 480-486 (1972).

Maurer, R.R., Whitener, R.H., Foote, R.H.: Relationship of in vivo gamete aging and exogenous hormones to early embryo development in rabbits. Proc. Soc. exp. Biol. Med. 131, 882-885 (1969).

Mittwoch, U.: DNA synthesis in cells grown in tissue culture from patients with mongolism. In: Mongolism Ciba Foundation Study Group N° 25, 51-61. London: Churchill 1967.

Mittwoch, U., Delhanty, J.D.A.: Inhibition of mitosis in human triploid cells. Nature (New Biol.) 238, 11-13 (1972).

Mittwoch, U., Delhanty, J.D.A., Beck, F.: Growth of differentiating testes and ovaries. Nature (Lond.) 224, 1323-1325 (1969).

Motlik, J., Fulka, J.: Fertilization and development in vivo of rabbit oocytes cultivated in vitro. J. Reprod. Fertil. 40, 183-187 (1974).

Niebuhr, E.: Triploidy in Man. Humangenetik 21, 103-125 (1974).

Nikawa, N., Kajii, T.: A triploid human abortus due to dispermy. Humangenetik 24, 261-264 (1974).

Niwa, K., Chang, M.C.: Fertilization in vitro of rat eggs as affected by the maturity of the females and the sperm concentration. J. Reprod. Fertil. 35, 577-580 (1973).

Niwa, K., Chang, M.C.: Optimal sperm concentration and minimal number of spermatozoa for fertilization in vitro of rat eggs. J. Reprod. Fertil. 40, 471-474 (1974).

Papiernik-Berkhauer, E.: Enfant triploïde né à terme et thérapeutique hormonale. Bull. Féd. Gynéc. Obstét. franç. 20, 248-250 (1968).

Paton, G.R., Silver, M.F., Allison, A.C.: Comparison of cell cycle time in normal and trisomic cells. Humangenetik 23, 173-182 (1974).

Piko, L.: La polyspermie chez les animaux. Ann. Biol. anim. Biochim. Biophys. 1, 323-383 (1961).

Piko, L., Bomsel-Helmreich, O.: Triploid rat embryos and other chromosomal deviants after cholchicine treatment and polyspermy. Nature (Lond.) 186, 737-738 (1960).

Pitkjanen, I.G.: Ovulation, fertilization and the early stages of embryonic development in pigs. Izv. Akad. Nauk Kazah. S.S.R., Ser. Biol. 31, 20-31 (1955).

Salisbury, G.W., Hart, R.G.: Gamete aging and its consequences. Biol. Reprod. Suppl. 2, 1-15 (1970).

Schmid, W., Vischer, D.: A malformed body with double aneuploidy and diploid-triploid mosaicism 48 XXYY/71 XXXYY. Cytogenetics 6, 145-155 (1967).

Schneider, E.L., Epstein, C.H.: Replication rate and lifespan of cultured fibroblasts in Down's syndrome. Proc. Soc. exp. Biol. Med. 141, 1092-1094 (1972).

Shaver, E.L., Carr, D.H.: Chromosome abnormalities in rabbit blastocyst following delayed fertilization. Reprod. Fertil. 14, 415-420 (1967).

Snow, M.H.L.: Tetraploid mouse embryos produced by Cytochalasin B during cleavage. Nature (Lond.) 244, 513-514 (1973).

Stranzinger, G.: Untersuchungen über den Einfluß der in vitro Spermaaufbewahrung auf die Befruchtung und den Fruchttod sowie die Möglichkeit der Analyse dadurch bedingter chromosomaler Veränderungen beim Kaninchen. Dissertation Univ. Göttingen 1970.

Tarkowski, A.K., Witkowska, A., Nowicka, J.: Experimental parthenogenesis in the mouse. Nature (Lond.) 226, 162-165 (1970).

Tesh, J.M.: Effect of time of insemination and superovulation on fertilization in the rabbit. J. Endocr. 35, XXVIII-IX (1966).

Thibault, C.: Analyse de la fécondation de l'oeuf de la truie après accouplement ou insémination artificielle. Ann. Zootechn. 8 (suppl.) 165-177 (1959).

Thibault, C.: Analyse comparée de la fécondation et de ses anomalies chez la brebis, la vache et la lapine. Ann. Biol. anim. Biochim. Biophys. 7, 5-23 (1967).

Vickers, A.D.: Delayed fertilization and chromosomal anomalies in mouse embryos. J. Reprod. Fertil. 20, 69-76 (1969).

Witkowska, A.: Parthenogenetic development of mouse embryos in vivo. I.Preimplantation development. II. Postimplantation development. J. Embryol. exp. Morphol. 59, 519-545 (1973).

Wroblewska, J.: Developmental anomaly in the mouse associated with triploidy. Cytogenetics 10, 199-207 (1971).

Yanagimachi, R., Chang, M.C.: In vitro fertilization of golden hamster ova. J. exp. Zool. 156, 361-369 (1964).

Autosomal Monosomy and Trisomy Causing Developmental Failure

A. GROPP, B. PUTZ, and U. ZIMMERMANN

I. Introduction

Systematic experimental investigations of the developmental consequences of autosomal monosomy and of trisomy seem to be an interesting task in view of the great clinical importance of chromosomal abnormalities in prenatal pathology of man (*Boué* et al., 1975; *Boué* and *Boué*, 1976). The establishment of preconditions for such studies has been possible by the observation that, in the mouse, structural heterozygosity for one or more metacentric chromosomes may enhance meiotic nondisjunction (*Gropp* et al., 1970; *Tettenborn* and *Gropp*, 1970; *Cattanach* and *Moseley*, 1973). Chromosomally hypo- or hypermodal gametes, which are responsible for (whole arm) monosomy or trisomy of the embryo, are produced as a consequence of irregular meiotic segregation.

Another experimental approach to studying autosomal trisomy in the mouse has been pointed out by *de Boer* (1973) who observed partial (so-called tertiary) trisomic animals in the offspring of the T70H translocation stock of the mouse. The affected animals are often viable and may be fertile. Therefore, they represent an interesting model for the study of meiosis and of reproduction in chromosomally unbalanced animals (*de Boer* and *Groen*, 1974). On the other hand, it is difficult in such tertiary trisomics to estimate and to determine the nature and the size of the trisomic chromosomal segment, so that their value for systematic studies of the developmental consequences of specific trisomic conditions is limited.

No other laboratory mammal besides the mouse is known in which autosomal monosomy or trisomy can be induced experimentally. However, spontaneous occurrence of autosomal trisomy has been observed occasionally in severely malformed calves (*Mori* et al., 1969) and, with a Down's-like syndrome of Ts 21, in nonhuman primates (*McClure* et al., 1969).

Sex chromosomal monosomy and trisomy, which are rather frequent in man and in animals (*Ford*, 1970), are often compatible with nearly normal somatic development (except the genital organs). They are not considered further in the present context, because this report is devoted to the discussion of autosomal monosomy and trisomy only.

II. Material and Cytogenetic Fundamentals of the Mouse Model

1. Animals

Metacentric chromosomes in the mouse occur as a result of casual processes of Robertsonian centric fusion of acrocentric chromosomes (Fig. 1), which otherwise represent the

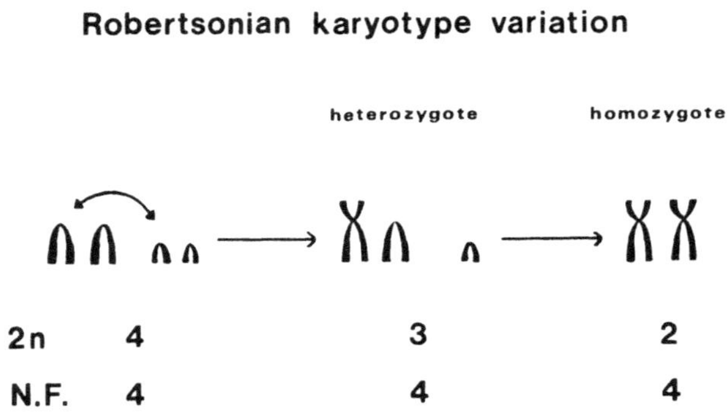

Fig. 1. Centric (Robertsonian) fusion causes decrease of chromosome number (2n), but leaves number of arms (N.F.) unchanged

normal type of chromosomes found in this species. The acrocentrics involved in the formation of a metacentric fully retain their genetic constitution, conserving, therefore, complete homology with the corresponding original acrocentric chromosomes in spite of the cytological rearrangement. A rather comprehensive reservoir of Robertsonian fusion metacentrics has been found particularly in feral populations (*Gropp* et al., 1972; *Capanna* et al., 1975). Mainly from this source, and in part from laboratory strains, 16 different metacentrics isolated and bred in single lines of the laboratory mouse (Table 1) are available so far for experimental use (*Léonard* and *Deknudt*, 1967; *White* and *Tjio*, 1967; *Baranov* and *Dyban*, 1971; *Cattanach* et al., 1972; own work, unpubl.). The cytogenetic individuality of a metacentric depends on the special selection of chromosomal arms involved in its formation. Partial monobrachial homology may exist between two different metacentrics. In this case, one of the two arms of the metacentrics is common to both.

Table 1. Mouse metacentric chromosomes

Designation [a] (arm composition)	Origin	Meiotic nondisjunction rates of male heterozygotes (Rb/+) in 1st meiotic anaphase [b]	
		$<20 + >20$ %	$>20 \times 2$ %
Laboratory strains			
Rb (15.6) 1Ald	*Léonard* and *Deknudt* (1967)	9	6
Rb (19.9) 163H	*Evans* et al. (1967)	5	4
Rb (19.5) 1Wh	*White* and *Tjio* (1967)		
Rb (17.8) 1IeM	*Baranov* and *Dyban* (1971)	8	6
Feral mouse populations			
Rb(3.1) 1 Bnr		15	14
Rb (6.4) 2Bnr		20	21
Rb (15.5) 3Bnr		27	28
Rb (13.11) 4Bnr		26	28
Rb (12.8) 5Bnr	*Gropp* et al. (1970, 1972)	7	4
Rb (14.9) 6 Bnr		11	10
Rb (17.16) 7Bnr		5	4
Rb (11.10) 8Bnr		6	4
Rb (12.4) 9Bnr		9	8
Rb (10.1) 10Bnr		4	4
Rb (15.4) 4Rma	*Capanna* et al. (1975)		
Rb (14.12) 7Rma			

[a] According to the COMMITTEE on STANDARDIZED GENETIC NOMENCLATURE for MICE (1972, 1974).

[b] On the basis of meiotic metaphase II arm counts (*Gropp* et al., 1975) and calculated according to *Cattanach* and *Moseley* (1973): Two estimates of nondisjunction frequency are listed, on the left total proportion of hypo- plus hypermodal MII cells (modal haploid count in the mouse: 20), on the right doubling of hypermodal MII. Since artifactual chromosome losses due to the cytological preparation can occur, the percentage on the left may overestimate, the one on the right may underestimate the true frequency (number of MII cells scored: between 300 and 500 per individual type of heterozygote).

2. Patterns of Meiotic Malsegregation (Nondisjunction) of Metacentric Heterozygotes

Mice heterozygous for Robertsonian fusion metacentrics may show considerable rates of malsegregation in the first meiotic division. There is evidence (*Gropp* et al., 1974) that such events concern in the first line the chromosomes involved in the formation of multivalents, which are characteristically found in meiosis I of metacentric heterozygotes. Thus, trivalents result from pairing of the two acrocentrics homologous with the arms of a metacentric in the case of single metacentric heterozygosity, while doubly metacentric heterozygous mice with one arm in common show a quadrivalent composed of both metacentrics and of two acrocentrics (Fig. 2).

Malsegregation in the first meiotic anaphase can be detected in the second meiotic metaphase (M II). Chromosome arm counts of M II plates of male single metacentric heterozygotes reveal different, in part remarkably high, rates of malsegregation (Table 1). Moreover, comparative studies of males versus females displaying the same type of chromosomal heterozygosity provided evidence that, almost regularly, nondisjunction rates are higher in the female than in the male sex (*Winking* and *Gropp*, 1976).

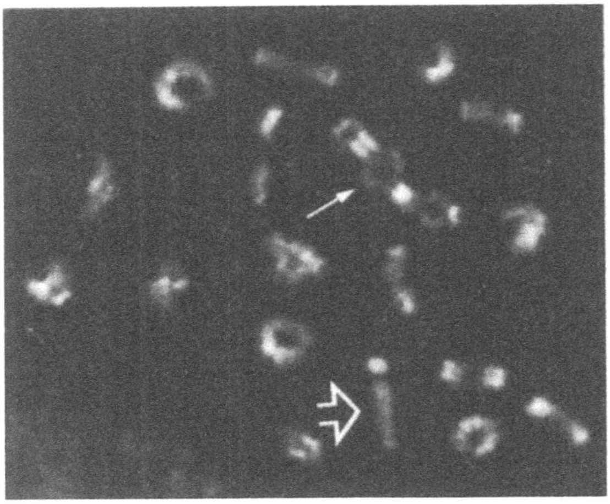

Fig. 2. Meiosis I (metaphase I) of double metacentric heterozygote Rb1/Rb10Bnr with quadrivalent (⟶) besides autosomal bivalents and X-Y-bivalent (⟸). 33258 H-fluorescence staining (*Gropp* et al., 1973)

Table 2.

a) Malsegregation rates of multiple metacentric heterozygotes in meiosis I of (tobacco mouse[a] × lab. mouse) F_1[c]

Rb1-7Bnr/+

♂ ♀

(n = 400) 52% ← chromosomally unbalanced (hypo- + hyperhaploid) MII → 68% (n = 100)

b) Survival pattern of euploid (chrom. arms = 40) and unbalanced (chrom. arms ≠ 40) embryos of (tobacco mouse[a] × lab. mouse[b]) F_1 ♂ and ♀ hybrids[c] backcrossed with lab. mice

Stage of development	Total no. of implants	No. of chromosomes (chrom. arms) hypodiploid ← balanced-diploid → hyperdiploid				
		39	39	40	41	41
		←		%		→
Preimplantation day 4-4 1/2	95	9	22	46	19	4
Day 8	128	–	2	63	24	11
Day 10-15	223	–	1	78	19	2
Day 20	58	–	–	97	3	–
Live born (1-3 days)	36	–	–	100	–	–

[a] Tobacco mouse (*M. poschiavinus*): 2n = 26 chrom., homozygous for 7 metacentrics Rb 1-7 Bnr

[b] Laboratory mouse: 2n = 40 ("all acrocentric" chrom). – Strain NMRI is used throughout the experiments

[c] F_1 hybrids: 2n = 33 chrom., heterozygous for Rb 1-7 Bnr

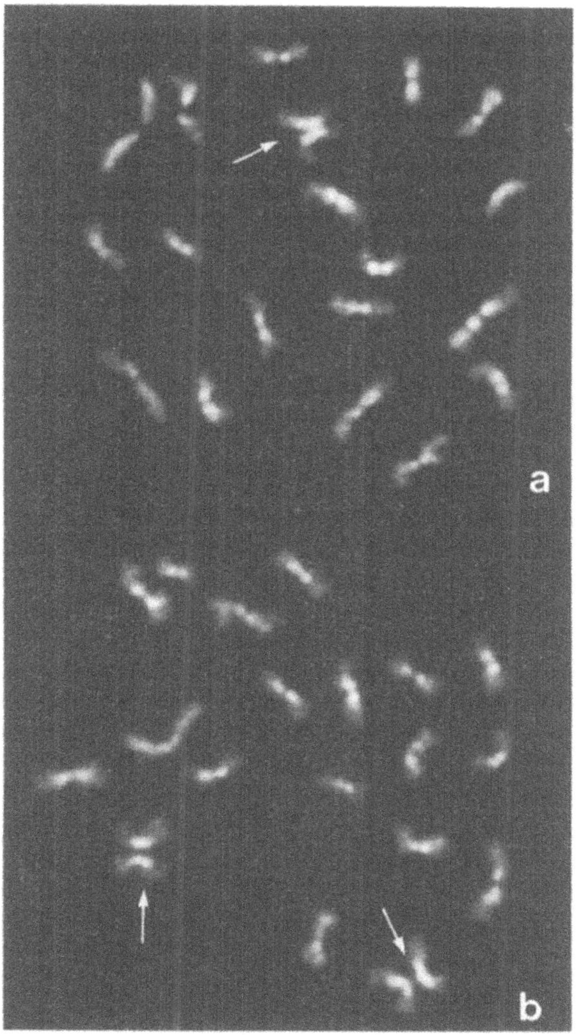

Fig. 3. Haploid metaphases of meiosis II (M II) – a) balanced chromosomal set with 20 chrom. arms in presence of one metacentric (→), b) hypermodal number of chrom. arms = 21 due to presence of two metacentrics (→). – 33258 H fluorescence (see Fig. 2)

Irregular segregation of the chromosomes involved in a trivalent leads to four types of unbalanced segregation products: namely two with hyperhaploidy due to an extra chromosome corresponding to either one of the two arms of the metacentric or, conversely, two types of hypohaploidy with lack of those chromosomes. With fertilization, mature gametes derived from hyperhaploid precursor cells will necessarily produce trisomy, and hypohaploid gametes, accordingly, monosomy of the zygote.

Cumulative rates of malsegregation occur in multiple heterozygotes with independent metacentrics (forming in meiosis independent trivalents) as, for example, in the F_1 generation of the feral tobacco mouse with seven metacentrics. These rates are again higher in females than in males (Table 2a).

Quadrivalents, on the other hand, as observed in double metacentric heterozygotes with partial homology (Fig. 2), are responsible for considerable frequencies of first meiotic anaphase nondisjunction. The respective numerical values are not only different according to the special double metacentric combination, but they are difficult to estimate (*Gropp* et al., 1975). They may virtually approximate to 50% or beyond. Unbalanced segregation products can either be hypohaploid, pseudohaploid, or hyperhaploid, possessing none, one, or two metacentrics (Fig. 3). However, the occurrence of the latter type with two metacentrics seems to be favored by a cytologically determined, though still obscure mechanism (*Winking* and *Gropp*, unpubl.). Hyperhaploid gametes derived from such gametogenic precursor cells whose metacentrics possess two copies of an homologous arm, by virtue of their monobrachial homology, are destined to contribute to the zygote an extra chromosome corresponding to this arm (see below).

III. Cumulative Data on the Incidence and Fate of Monosomy and Trisomy in Fetal Mice

Through studying the fetal backcross progeny of tobacco mouse F_1 hybrids (*Gropp* et al., 1970), an orientating survey of the role and incidence of whole arm autosomal aberrations produced by metacentric heterozygotes, monosomy as well as trisomy, is possible. Due to the presence of as many as seven metacentrics (Rb1-7Bnr), this system of a multiple metacentric heterozygote induces a sufficiently high rate of meiotic malsegregation (Table 2a) and thus, of zygote aneuploidy (Table 2b). Chromosome analyses show that in the preimplantation period, considerable proportions of hypomodal (monosomic) and hypermodal (trisomic) zygotes are observed by approximation with symmetrical distribution around the peak value of the modal (diploid) embryos. The apparent overrepresentation of hypomodal values is the result of artifactual chromosome losses caused by the chromosome preparations of blastocysts. Subsequently, during early and midterm embryonic stages, the monosomics disappear very rapidly whereas trisomic embryos survive longer, although they too become eliminated until the end of pregnancy. But, with the overall range of the developmental profiles, attempts of analyzing details of individual trisomic conditions of the mouse fetus seem to be a significant task.

IV. Breeding Designs for Inducing Specific Trisomics

Several designs of breeding animals heterozygous for metacentric chromosomes are suitable for inducing trisomy in their progeny: a) backcross of single metacentric heterozygotes with the "all acrocentric" laboratory mouse (see p. 181; *Gropp* et al., 1974), b) use of double metacentric heterozygotes with monobrachial homology which may contribute selectively individual extra chromosomes to a zygote in crosses with "all acrocentric" mice (see above; *White* et al., 1974; *Gropp* et al., 1975), and c) crosses between double metacentric heterozygotes inter se or between double metacentric heterozygotes and metacentric homozygotes (*White* et al., 1974). In all these cases, monosomies and trisomies as well as, in b and c, combined monosomy and trisomy, can be expected in the fetal progeny, but according to the observations in the "multiple heterozygote system" (Table 2b), trisomies are the only unbalanced offspring which survive after day 10.

Our own attempts at a systematic study of trisomy in the mouse fetus are based on outcrossing monobrachially homologous double metacentric heterozygotes with normal laboratory mice (see above b). Using this model, the occurrence of a particular and specific trisomy can be predetermined by the selection of the special double heterozygote combination, because the extra chromosome contributed by such combination to a trisomic condition corresponds to the arm shared by both metacentrics (see p. 182; Fig. 4). With the presently available metacentrics listed in Table 1, 12 out of the possible 19 autosomal trisomic conditions in the mouse can be induced, namely Ts 1, 4, 6, 8, 9, 10, 11, 12 (Fig. 4), 14, 15, 17, and 19.

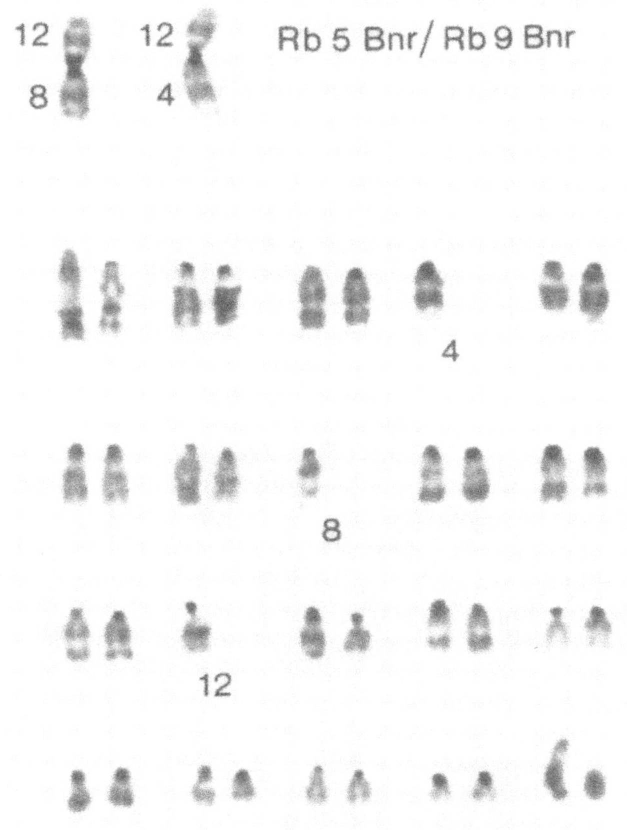

Fig. 4. Karyotype of mouse embryo with Ts 12. Triplication of chrom. no. 12 is due to one element (4th row) of nonheterozygous parent and to two copies contributed by metacentrics (1st row) of heterozygous parent. Direct preparation from fetal membranes. Giemsa-banding

The relative proportion of trisomic implants in the fetal progeny of double metacentric heterozygotes varies according to the individual trisomic conditions and it depends on whether males or females are used as heterozygous progenitors. But the actual frequency of trisomic embryos must be considered mainly against the background of the developmental profile, the life span, and the sequential elimination resulting from the mortality rates of the particular trisomic condition. Thus, to give examples for trisomies in the progeny of male heterozygotes, the mean frequency between day 10-14 is 16-17% for Ts 1, 8% for Ts 11, and 14.5% for Ts 12 (*Gropp*, 1975). In the case of Ts 19, *White* et al. (1974) refer to incidences of 27% (progeny of male heterozygote) and 8% (progeny of female heterozygote) between day 12 1/2 and 15 1/2.

Our own experiments are designed according to the following scheme: Mating of double metacentric heterozygous male (preferably) or female with NMRI. Demonstration of vaginal plug determines day 1 of development. Sacrifice of pregnant females at days 10, 12, 13, 14, and 15. Record of corpora lutea, total number of implants, number of males, living normal, or malformed embryos. Direct chromosome preparations from the fetal membranes of all living embryos and use of the intact embryo for embryologic studies.

V. Developmental Characteristics of Trisomy in the Mouse Fetus

The developmental profiles of autosomal trisomy in the mouse differ considerably according to which individual autosome is involved. In addition, it seems to be a common feature of specific trisomic conditions, that great variations of phenotype expression may occur (e.g., Fig. 5). This phenomenon is apparently due to the allelic heterogeneity of mouse stocks to which the introduction of metacentrics is – at present – limited. As mentioned before, trisomy in the mouse is almost invariably connected with earlier or later prenatal death (Table 2), though it is claimed by *White* et al. (1974) and *Baranov* (1974) that in

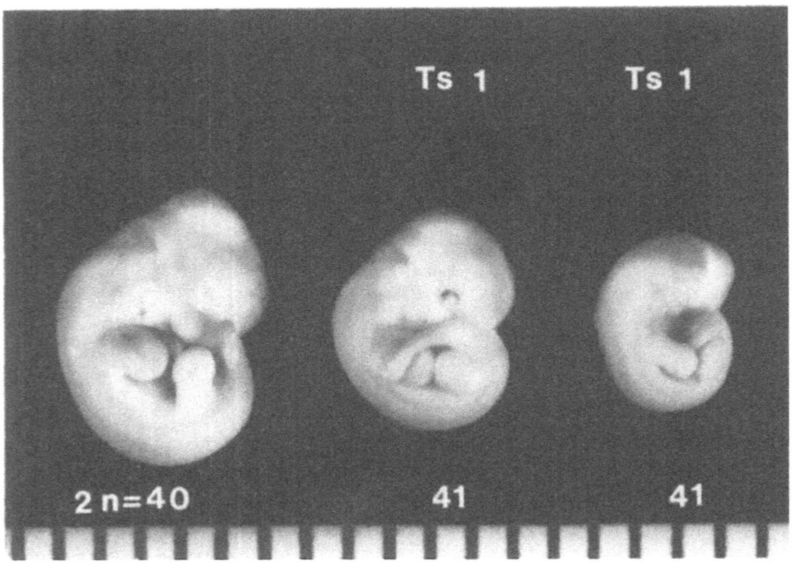

Fig. 5. Trisomy 1 (b, c) together with normal "litter mate" (a). Facial hypoplasia in heavier affected trisomic embryo. Fetal progeny of Rb1/Rb10Bnr heterozygote; day 12. mm-scale

certain trisomic conditions (Ts 19; Ts 14?) brief postnatal survival may occur. From a general point of view, there is no doubt that these features are very similar to those governing the fate of trisomy in man, although in this latter case limited postnatal survival is possible in certain specific trisomies (*Boué* and *Boué*, 1976).

1. Morphologic Features of Certain Individual Trisomies

In the mouse, some specific trisomies are characterized by severe hypoplasia, extreme developmental retardation or inhibition, and comparatively early death around day 11 or 12. Thus, Ts 4, 8, 11, 15, and 17 usually form only a minute unorganized mass or an extremely retarded embryo.

Ts 6 and Ts 10 of the fetal mouse display a syndrome of slight to moderate, sometimes severe retardation and hypoplasia. The developmental impairment and the variation of individual phenotype expression may range from near normality to marked hypoplasia. Usually, the placenta is also affected, though slightly, and shows concomitant changes of hypoplasia and retardation.

More defined malformation syndromes have been observed in Ts 19 described by *White* et al. (1974), and Ts 1 and 12 (*Gropp* et al., 1974, 1975).

Ts 19 can be associated with isolated cleft palate if a special chromosomal "background" involving three metacentrics is present, while low weight of embryo and placenta is a more general feature of this trisomy, in which limited postnatal survival seems to be possible, at least in some instances. The observation that cleft palate occurs only under special cytogenetic conditions demonstrates clearly the effects of the genic background upon the morphologic expression of trisomy.

In trisomy 1, in which the phenotype very often shows a remarkable range of variation, the more severely affected embryos exhibit smallness already on day 12 and 13 (see Fig. 9) and, in particular, craniofacial hypoplasia. Figure 5 demonstrates such differences of Ts 1 phenotype expression on day 12. But heavier craniofacial dysmorphy is found more frequently with further fetal development until day 15, which the embryos with Ts 1 hardly survive. The real nature of the developmental anomaly of Ts 1 is better understood from histological examination of the head which, in the severely affected embryos, reveals the impairment of the development of the prosencephalon: In contrast to normal embryos, the ocular primordia are small or even absent and the primitive forebrain vesicle is incompletely subdivided in a diencephalic and telencephalic area corresponding to earlier developmental stages of holoprosencephaly and cyclopia. In fact, some specimen of Ts 1 have been observed exhibiting almost classical features of cyclopia with proboscis, of which Figure 6 shows an example at day 15.

A parallel can be drawn with the Patau syndrome in man, which is similarly caused by an endogenous chromosomal anomaly, namely trisomy 13. Cyclopia represents, also in man, the most severe manifestation of Ts 13, while milder forms show only minor malformations like microcephaly, defect of the corpus callosum, etc.

As a third type of trisomy with specific fetal malformation, Ts 12 exhibits rather uniformly, and without exception, exencephaly (Fig. 7) and microphthalmia (enophthalmia), but

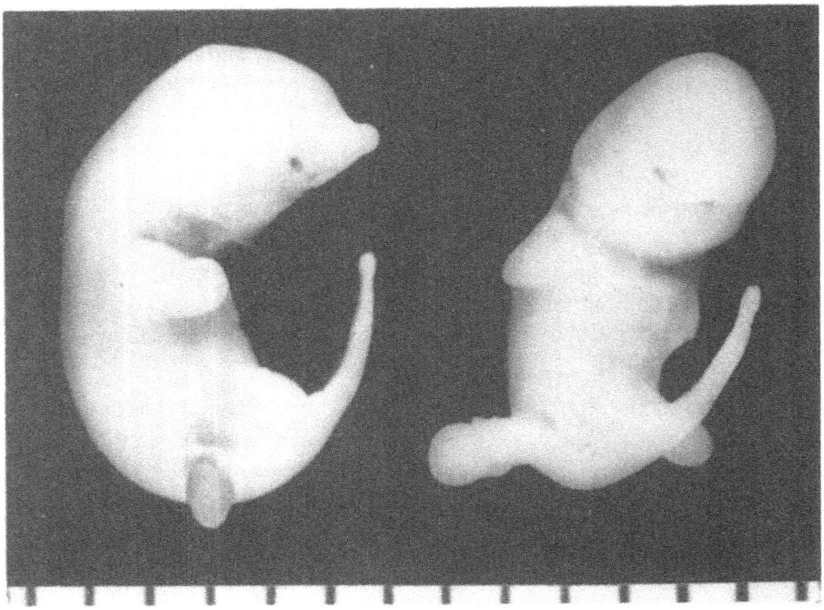

Fig. 6. Embryo with Ts 1, lateral and front view. Holoprosencephaly and cyclopic malformation. Day 15. mm-scale

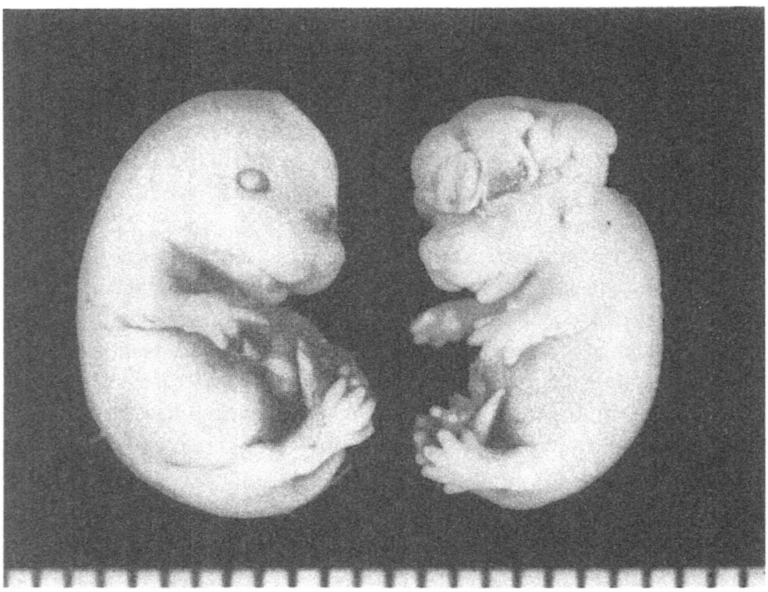

Fig. 7. Embryo with exencephaly due to Ts 12 (b) together with normal embryo (a) from same fetal litter of Rb5/Rb9Bnr heterozygote. Day 16 1/2. mm-scale

only slight to moderate hypoplasia or retardation (see Fig. 9). Death of these embryos occurs between day 15 to 17. Histologically, the brain is extroverted, though mostly quasi-normal with respect to the morphogenetic development of its parts. Only in a minority of embryos, more severe impairment of the development of the brain similar to anencephaly is observed. Caudal neural tube defects, in particular spina bifida, are absent in Ts 12. Histologic examination (Fig. 8) reveals a marked hypoplasia of the cartilagenous skull base in

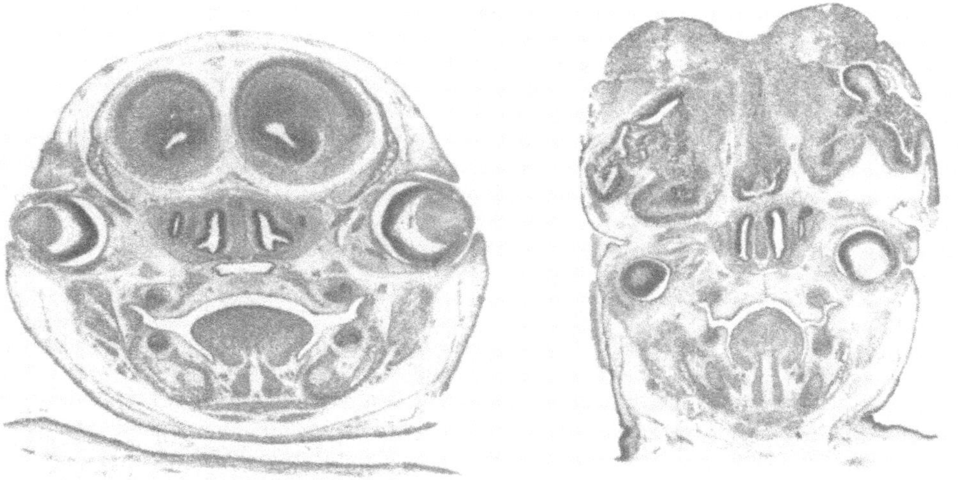

Fig. 8. Frontal histolocial section of normal embryo (a) and of embryo with Ts 12 showing exencephaly (b). Developmental stage day 16

trisomic embryos. Detailed morphometric and stereologic measurements (*Putz*, unpubl.) show an abnormal lateral downward inclination of the basisphenoid, which entails a similar downward move of the angle of the axis of the optic nerve and ocular bulbs. So far, these changes are similar to those described in exencephaly induced by exogenous agents, e.g., vitamin A, but in the Ts 12 syndrome they are responsible for an inward displacement of the eye bulb suggesting microphthalmia. Conversely, in the exogenously (vitamin A) induced syndrome, exophthalmia may occur due to an outward move of the ocular bulbs in consequence of more fundamental concomitant textural changes of the facial primordia (*Putz* and *Gropp*, unpublished observations) absent in Ts 12. These changes, which are responsible for a size reduction of the orbita in the exogenous exencephaly syndrome, affect most heavily the region of the oropharynx. With respect to these differences, the comparable malformations in man, mainly anencephaly, resemble more closely the situation in Ts 12.

2. General Aspects of Developmental Impairment in Trisomy

In the syndromes of Ts 1 and Ts 12 of the fetal mouse, the target area of the teratological effect of the chromosomal abnormality is probably restricted to tissular units of induction in the development of the basis of the skull and anterior neural tube, that is, mainly to the prechordal plate, and possibly to a specific notochordal constituent of this plate. This assumption is in accordance with the fact that the malformations observed in Ts 1 and Ts 12 (holoprosencephaly and exencephaly) are strictly limited to the brain and the cerebrocranium, whereas in the exogenously (e.g., vitamin A) induced syndrome of exencephaly, the damage affects much wider areas, including also migrating constituents of the neural crest and segments of the caudal notochord as well (*Kalter* and *Warkany*, 1961; *Langman* and *Welch*, 1966). The effect of trisomy leading to a defective induction is probably confined to about day 8 1/2 in Ts 12, and to a slightly later period around day 9 to 9 1/2 in Ts 1. In the case of Ts 1, the degree and completeness of the inductive action impairment determine the severity of the malformation complex, which shows, in fact, a considerable variation of expression, with holoprosencephaly and cyclopia at the end of the scale.

As almost all the malformations observed so far in trisomy concern the head region, it seems noteworthy that limb malformations have not yet been found.

One would, of course, expect the triplication of a chromosome, which carries a great number of genes, to cause a severe derangement of the regulation within a multigenic system. Therefore, the phenotype of trisomy is probably determined more by a pattern of indirect consequences of genome unbalance, than by direct effects of gene actions. The finding that the malformations found in trisomic conditions correspond mainly to inductional disorders, is in full agreement with this assumption. On the other hand, it is doubtful whether malformation per se can explain the development failure of trisomy. Instead, it is likely that gross malformation indicates a more local damage of specialized blastemas only, for example local growth disorders, or a defect of induction and conditioned differentiation, which develop when the embryos live long enough. This could result, in particular, in blastemas and organs entering specific phases of cellular proliferation and differentiation with a relative deficiency of cells (*Polani*, 1974). An early breakdown of growth and differentiation, on the other hand, is observed as a more severe phenotype expression of the triplication of certain autosomes, and even more in the case of haplozygosity of a chromosome as in the autosomal monosomies.

Hence, it can be assumed that the failure of development of aneuploid embryos is, to a large extent, due to impairment and slowing down of growth leading eventually to death. One of the main features of the development of trisomic embryos is the lag of growth which, however, might be of varying severity depending on the type of trisomy. Figure 9 demonstrates these facts showing delayed increase in weight of embryos with Ts 1 and Ts 12, compared with normal embryos.

According to preliminary results (*Citoler* et al., unpubl.) of comparative studies on the in vitro growth patterns of fetal cells from Ts 1, Ts 12, and normal mouse embryos, it appears that no clear-cut changes of cell cycle kinetics can be found in in vitro cultures of trisomic tissues. Moreover, the proportion of cells participating in the growth fraction does not differ between normal and trisomic cells. But, a slight to moderate impairment of the overall growth and doubling time, and a more definite decrease of the life span can be noted in trisomic monolayer cultures, whereas no major size differences of the growth areas of con-

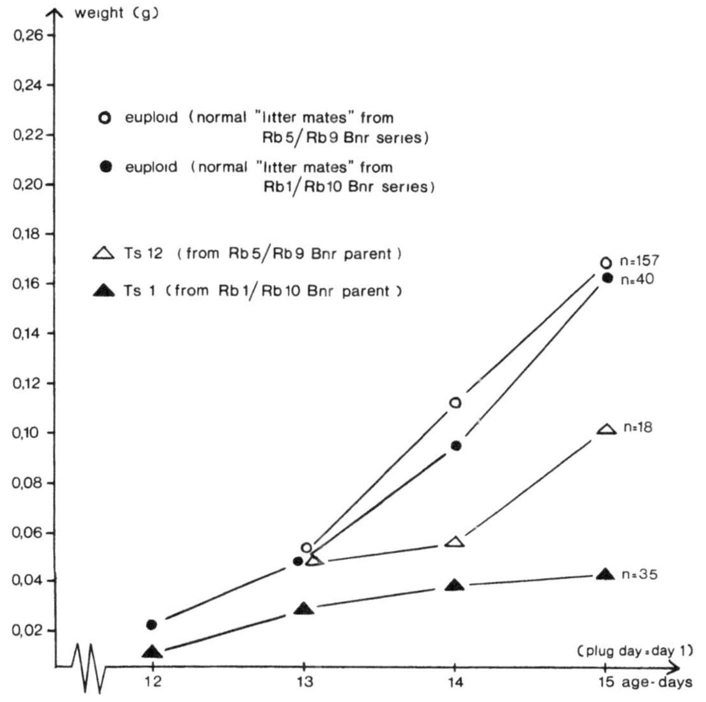

Fig. 9. Weight increase of normal and trisomic embryos. n = total number of embryos weighted per category. Retarded increase in Ts 12, and more complete lag of weight increase in Ts 1

fronted normal and trisomic solid explants grown together in the same culture vessel were observed. In this latter case, it is, however, necessary to exclude the effect of complementation between normal and trisomic cells. The attempts to evaluate the growth characteristics and the growth kinetics of trisomic cells in vitro could not yet elucidate the nature and mechanism of the growth lag and of the developmental retardation which certainly represent the main features of trisomy in vivo. It remains to be seen whether further systematic studies in trisomy in the mouse are able to provide a better insight and more detailed information, permitting the establishment of a unifying concept for some descrepant observations on cell cycle changes (*Kuliev* et al., 1973; *Paton* et al., 1974) and on the decrease of the growth rate and reduced life span in vitro as described by several authors in cases of trisomy in man (*Schneider* and *Epstein*, 1972; *Cure* et al., 1974; for review: *Boué* and *Boué*, 1976).

Finally, the observation that the trisomic placenta is similarly involved in the impairment of growth and differentiation, raises the question as to whether a disproportionate manifestation of such effects on the placenta could initiate the developmental breakdown of the embryo, for example, by means of an impaired permeability for essential metabolites, or, by means of insufficiency of the fetal blood supply. The evaluation of the incorporation of tritium labeled serine, phenylalanine, and thymidine in the placenta and in the embryo on the background of measurements of the free compounds and of dry weight of these tissues lead to the conclusion (*P. Citoler* et al., 1975) that there is no impairment of the permeability of the trisomic placenta, at least not for the two amino acids and for thymidine. The observation that the actual incorporation of both amino acids in trisomic embryos is decreased at variable rates can be interpreted as another expression of the impairment of cellular growth-related metabolic activities. There is, however, certain histomorphologic evidence of incomplete morphoarchitectural differentiation of trisomic placentas, mainly in the zona spongiosa of the decidua basalis and the labyrinth (*Zimmermann*, unpubl.). These latter findings need further examination with respect to their functional significance.

VI. Conclusion and Summary

A mouse model is described which allows the experimental study of fetal monosomy and trisomy. By using a particular design of breeding doubly metacentric heterozygous animals, specific trisomies can be induced systematically. Considering the information presently available on the growth and developmental patterns of monosomy and trisomy, a general picture emerges from the mouse model which is characterized by a sequential elimination of the chromosomally abnormal progeny depending on the special chromosome involved. An early selection operates against monosomics, whereas in trisomy, breakdown of growth and development occurs later. Gross malformation, apparently with prevalence of inductional disorders in the head region, develops mainly in later stages. It appears that the animal model in the mouse reflects perfectly the sequence of events observed in zygotic and prenatal chromosomal disorders in general, and that it can be used for the elucidation of the factors governing prenatal developmental disorders also in man.

Acknowledgments. The Deutsche Forschungsgemeinschaft (Pränatale Diagnose) supported this work. We thank Mrs. *U. Kolbus* for her great help.

References

Baranov, V.S.: Trisomy in the newborn. Experiments on mice heterozygous for the T_6 Bnr translocation. Bull. exp. Biol. Med. 5, 99-103 (1974).

Baranov, V.S., Dyban, A.P.: Embryogenesis and pecularities of karyotype in mouse embryos and centric fusion of chromosomes (Robertsonian translocation). Ontogenese Akad. Nauk SSSR 2, 164-175 (1971).

Boer, P. de: Fertile tertiary trisomy in the mouse (Mus musculus). Cytogenet. Cell Genet. 12, 435-442 (1973).

Boer, P. de, Groen, A.: Fertility and meiotic behavior of male T 70 H tertiary trisomics of the mouse (Mus musculus). Cytogenet. Cell Genet. 13, 489-510 (1974).

Boué, J.G., Boué, A.: Chromosomal anomalies in early spontaneous abortion. Current Topics in Pathology (ed. *A. Gropp*). 62, 193-208 (1976).

Boué, J., Boué, A., Lazar, P.: The epidemiology of human spontaneous abortions with chromosomal anomalies. Aging Gametes, Int. Symp., Seattle 1973, pp. 330-348. Basel: Karger 1975.

Capanna, E., Cristaldi, M., Perticone, P., Rizzoni, M.: Identification of chromosomes involved in the nine Robertsonian fusions of the Appenine mouse with a 22-chromosome karyotype. Experientia (Basel) 31, 294-296 (1975).

Cattanach, B.M., Williams, C.E., Bailey, H.: Identification of the linkage groups carried by the metacentric chromosomes of the tobacco mouse (Mus poschiavinus). Cytogenet. Cell Genet. 11, 412-423 (1972).

Cattanach, B.M., Moseley, H.: Nondisjunction and reduced fertility caused by the tobacco mouse metacentric chromosomes. Cytogenet. Cell Genet. 12, 264-287 (1973).

Citoler, P., Gropp, A., Kolbus, M., Reinhardt, P., Schaefer, R.: Metabolism of the placenta and foetus in the trisomy-syndrome of mice. IV. Konferenz der Europäischen Gesellschaft für Teratologie, Freudenstadt 23.-25. Sept. 1975. Abstract.

Cure, S., Boué, A., Boué, J.: Consequence of chromosomal anomalies on cell multiplication. Symposium "Les accidents chromosomiques de la réproduction" (ed. *A. Boué, Ch. Thibault*), pp. 95-109. INSERM 1973.

Ford, C.E.: The population cytogenetics of other mammalian species. Pfizer Med. Monogr. 5: Human Population Cytogenetics, pp. 222-239. Edinburgh University Press 1970.

Gropp, A.: Chromosomal animal model of human disease. Fetal trisomy and developmental failure. In: Teratology (ed. *L. Berry, D.E. Poswillo*), pp. 17-33. Berlin–Heidelberg–New York: Springer 1975.

Gropp, A., Giers, D., Kolbus, U.: Trisomy in the fetal backcross progeny of male and female metacentric heterozygotes of the mouse. I. Cytogenet. Cell Genet. 13, 511-535 (1974).

Gropp, A., Hilwig, I., Seth, P.K.: Fluorescence chromosome banding patterns – produced by a benzimidazole derivative. Proc. 23rd Nobel Symposium. Stockholm, September 25-27, 1972 (ed. *T. Caspersson, L. Zech*), pp. 300-306. The Nobel Foundation, Stockholm and Academic Press, New York 1973.

Gropp, A., Kolbus, U., Giers, D.: Systematic approach to the study of trisomy in the mouse II. Cytogenet. Cell Genet. 14, 42-62 (1975).

Gropp, A., Tettenborn, U., Lehman, E. von: Chromosomenvariation vom Robertson'schen Typus bei der Tabakmaus M. poschiavinus und ihren Hybriden mit der Laboratoriumsmaus. Cytogenet. Cell Genet. 9, 9-23 (1970).

Gropp, A., Winking, H., Zech, L., Müller, H.J.: Robertsonian chromosomal variation and identification of metacentric chromosomes in feral mice. Chromosoma (Berl.) 39, 265-288 (1972).

Kalter, H., Warkany, J.: Experimental production of congenital malformations in strains of inbred mice by maternal treatment with hypervitaminosis A. Amer.J. Path. 38, 1-21 (1961).

Kuliev, A.M., Kukharenko, V.I., Grinberg, K.N.: Morphological autoradiographic, immunochemical and cytochemical investigation of a cell strain with trisomy 7 from a spontaneous abortus. Humangenetik 17, 285-296 (1973).

Langman, J., Welch, G.W.: Effect of vitamin A on the development of the central nervous system. J. comp. Neurol. 128, 1-15 (1966).

Léonardt, A., Deknudt, G.H.: A new marker for chromosome studies in the mouse. Nature (Lond.) 214, 504-505 (1967).

McClure, H.M., Belden, K.H., Pieper, W.A., Jacobson, C.B.: Autosomal trisomy in a chimpanzee: resemblance to Down's syndrome. Science 165, 1010-1012 (1969).

Mori, M., Sasaki, M., Makino, S., Ishikawa, T., Kawata, K.: Autosomal trisomy in a malformed newborn calf. Proc. jap. Acad. 45, 955-959 (1969).

Paton, G.R., Silver, M.F., Allison, A.C.: Comparison of cell cycle time in normal and trisomic cells. Humangenetik 23, 173-182 (1974).

Polani, P.E.: Chromosomal and other genetic influences on birth weight variation. In: Size at Birth, Ciba Found. Symp. 27 (new series), pp. 127-164. North Holland: Elsevier Exc. Medica 1974.

Schneider, E.L., Epstein, Ch.J.: Replication rate and life span of culture fibroblasts in Down's syndrome (36940). Proc. Soc. exp. Biol. (N.Y.) 141, 1092-1094 (1972).

Tettenborn, U., Gropp, A.: Meiotic nondisjunction in mice and mouse hybrids. Cytogenetics 9, 272-283 (1970).

Winking, H., Gropp, A.: Meiotic non-disjunction of metacentric heterozygotes in oocytes versus spermatocytes. Serono Symp. Freiburg, 10.-12. Sept. 1975. New York–London: Academic Press, in press.

Chromosomal Anomalies in Early Spontaneous Abortion

(Their Consequences on Early Embryogenesis and in vitro Growth of Embryonic Cells)

J.G. BOUÉ and A. BOUÉ

During the first half of the century anatomical studies of human spontaneous abortions have demonstrated the importance of abnormalities in the development of the conceptus, and, beginning in 1963, various studies have demonstrated the high frequency of chromosomal anomalies in these abortuses.

Cytogenetic analysis of a large number of abortuses, coupled with anatomical examination and tissue culture studies, have made it possible to establish correlations between these chromosomal anomalies and (1) their consequences on early embryogenesis and (2) in vitro growth characteristics of embryonic cells which carry these anomalies.

I. Frequency of Chromosomal Anomalies in Early Arrests of Embryogenesis

The abortions were first classified according to the stage of development. In clinical studies the length of gestation is commonly estimated from the date of the last menstrual period, but the gestational age is an extremely invalid means of determining the developmental stage of the zygote. This duration is in fact determined by maternal rather than zygotic factors. Menstrual age represents the summation of three periods: (1) the preconceptional period arbitrarily taken as 14 days; (2) the length of the zygotic development, and (3) the period of in utero retention between the arrest of zygotic development and expulsion of the product of conception. An analysis of the duration of gestation in 716 abortions showed that the average length of in utero retention is between 6 and 8 weeks. Furthermore, hormonal treatment can modify the length of retention: after progesterone treatment the mean length of in utero retention was 7.9 weeks, whereas the control group of untreated women had a retention of 5.9 weeks.

Anatomical studies have demonstrated that the duration of development is a good criterion for the classification of specimens. An estimation of the stage of development attained by the zygote may be based on: a) morphologic examination of the embryo when an embryonic formation exists; the age of the embryo is estimated from the stage of embryogenesis it has attained rather than from its size, which is usually modified by its state of maceration, and b) detailed histologic examination of the placenta including an appreciation of the degree of maturity of the villi and their blood vessels. Thus, correlations can be established using criteria directly related to the conceptus.

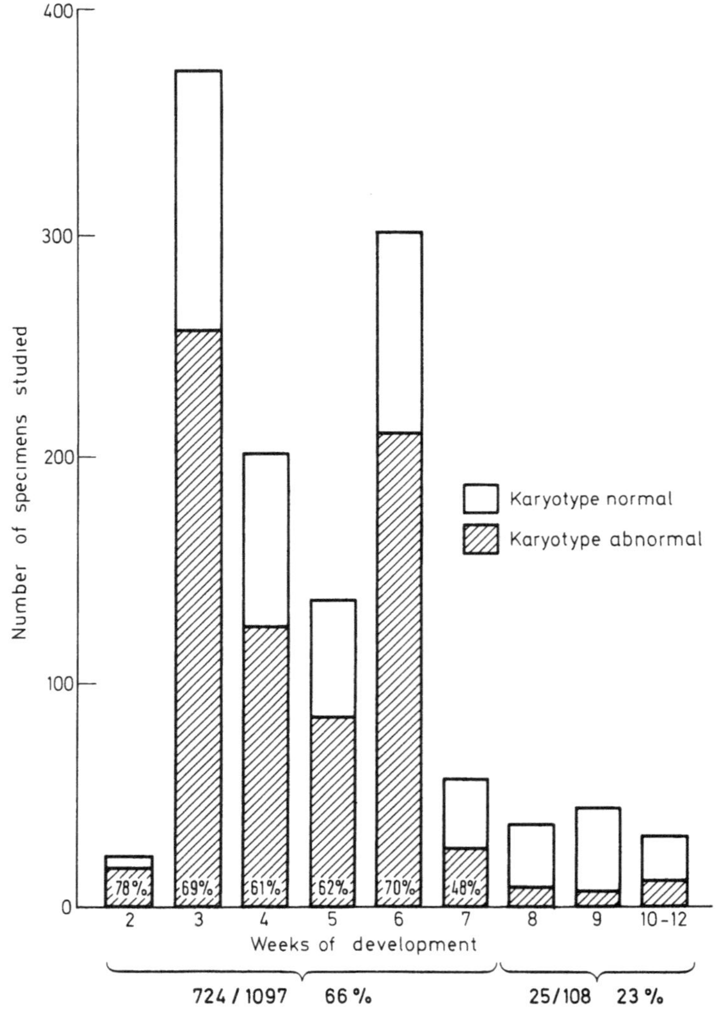

Fig. 1. Frequency of chromosomal anomalies in abortuses in relation to developmental arrest (*Boué* and *Boué*, 1974)

It has been possible to establish the date of developmental arrest in 1205 abortions. Figure 1 shows the frequency of chromosomal anomalies according to the development attained by the zygote. In this series only zygotes with a development of less than 12 weeks (postconception) were studied. Nine times out of ten (1097/1205 observations) the developmental arrest occurred before the eighth week. Developmental arrests before the third week are rare because in these cases the pregnancy is usually unrecognized and such specimens are seldom received for examination. The frequency of anomalies is about 66% at 3-7 weeks, and falls to 23% for zygotes with 8-12 weeks development.

II. Types of Chromosomal Anomalies

Table 1 shows the results of chromosomal analysis of the 1498 abortuses studied: 921 had chromosome anomalies.

Table 1. Karyotypes of 921 abortuses with abnormal karyotype

		Number		Percentage
Monosomy	45 X	140	141	15.30
	45, G –	1		
Trisomy	A +	12	479	52.00
	B +	6		
	C +	86		
	D +	109		
	E +	172		
	F +	7		
	G +	87		
Double trisomy		16		1.73
Triploidy	XXY	92	183	19.86
	XYY	7		
	XXX	57		
	Unkaryotyped	27		
Tetraploidy		57		6.18
Translocations		35		3.80
Mosaicism		10		1.08

Structural anomalies were observed in 3.8% of abortuses with anomalies. In only one-third of these observations was the anomaly transmitted by one of the parents. Thus, nearly all chromosomal aberrations observed in abortuses are the result of errors which occurred during gametogenesis or at the time of fertilization.

Chromosomal nondisjunction during meiosis (male or female) may lead to monosomic or trisomic zygotes.

Monosomy X (45,X) represents one of the most frequently encountered anomalies in spontaneous abortions, 15% of the total in our study. Autosomal monosomies are extremely rare; a few isolated cases have been reported.

Spontaneous abortions with a trisomic karyotype represent more than 50% of the anomalies. In contrast to the monosomies in which the missing chromosome is almost always a sex chromosome, in trisomies the extra chromosome is nearly always an autosome. In abortions, trisomies of all chromosome groups are found, but their relative frequencies vary: trisomy C, D, E, and G are frequent, whereas trisomy A, B, and F are rare.

Since this study was carried out before the banding techniques became available it has not always been possible to distinguish which chromosome of a given group (group C for example) was additional. In good preparations, however, the morphology of some chromosomes is sufficiently characteristic for their precise identification; thus trisomy 16 is clearly recognizable and is very frequent. It accounts for 75% of the E trisomies and amounts to nearly 15% of all chromosomal anomalies observed in abortuses.

During this study, attempts to initiate cell lines from embryonic tissues were carried out as often as possible and these cell lines were stored frozen. Recently an identification of the extra chromosome in the trisomic cells has been accomplished with the banding techniques, demonstrating the occurrence of trisomies 7, 8, 9, 10, 12, 13, 14, 15, 16, 18, 20, 21, and 22.

Triploidy accounts for about 20% of the anomalies found in abortions. Triploidy can be explained by either diandry or dispermy with two sets of paternal chromosomes and one maternal set, or digyny, with one paternal set and two maternal sets. Using different approaches, an attempt has been made to ascertain whether these two mechanisms occur in human beings and to determine their relative frequency. The fact that 69,XYY karyotypes have been observed proves that diandry occurs. The small number of abortuses with 69,XYY constitution does not necessarily reflect the infrequency of this mechanism, which could be explained by the fact that the 69,XYY karyotype leads to a very early developmental arrest. These cells do not grow very well in tissue culture, and we have never been able to initiate cell lines from cells with 69,XYY karyotypes.

Using fluorescent chromosome markers, diandry and digyny have been demonstrated (*Jonasson* et al., 1972; *Uchida* et al., 1972). In a study in progress with Jacques Hors and Jean Dausset, the expression of HL-A antigens on some triploid cell lines initiated from embryos has been determined and correlated with the HL-A antigens of the parents. In the following example: triploid abortus, 69,XXY sex chromatin negative, the father was HL-A1,W28, HL-A8,W5; the mother was HL-A2,D25, W10, W15. In the triploid cells, six antigens were identified: HL-A1, W28, D25, HL-A8,W5, and W10. The presence of the four HL-A antigens of the father clearly demonstrates the mechanism of dispermy.

Tetraploidy represents about 5% of the anomalies, 92,XXXX being the most common karyotype (33 in the 53 tetraploidies in our series). The fact that in all the observations the tetraploid abortuses have an XXXX or XXYY sex chromosome constitution supports the mechanism of a failure of cytokinesis during the first division of the zygote. One observation of a zygote with 94 chromosomes, 94, XXXX, 16+, 16+, is explained by this mechanism occurring in a zygote with trisomy 16.

III. Evaluation of Chromosomal Errors at Conception

The relative frequency of the different types of chromosomal anomalies in early arrests of embryogenesis is similar in the different surveys published. Figure 2 shows the results of four series: *Boué* (France), *Carr* (Canada), *Kajii* (Switzerland), and *Therkelsen* (Denmark). All these studies have confirmed the high frequency of monosomy X and trisomy 16, each representing at least 15% of the anomalies observed.

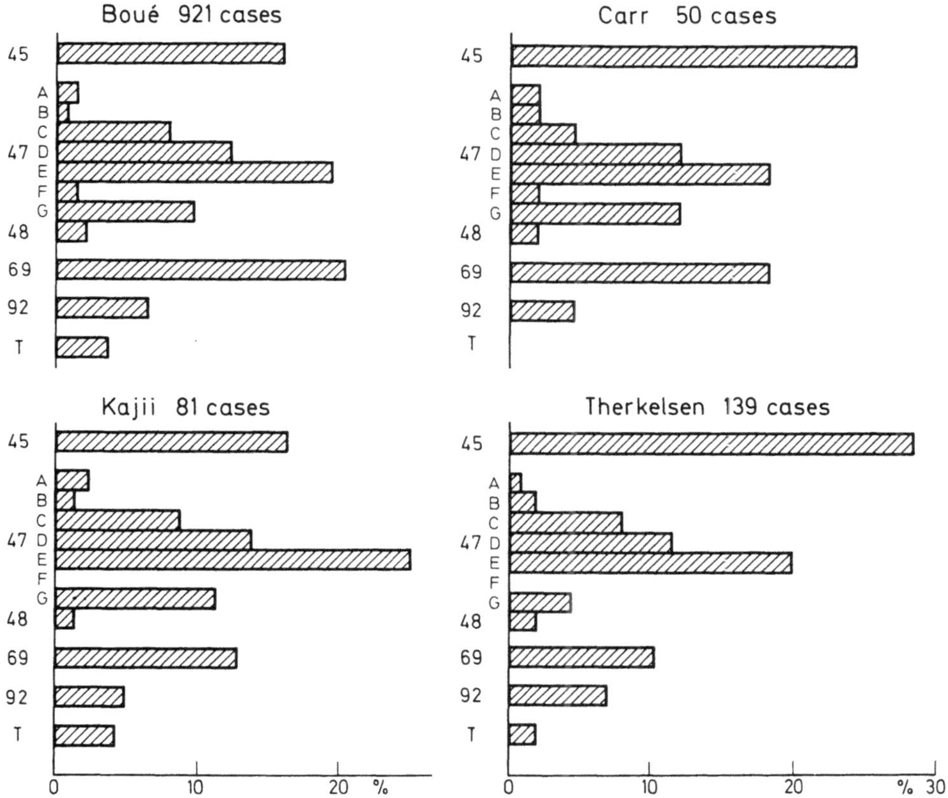

Fig. 2. Relative frequency of different chromosomal anomalies in spontaneous abortions. Results from *Boué* and *Boué* (1974), *Carr* (1967), *Kajii* et al. (1973) and *Therkelsen* et al. (1973)

To account for the absence of autosomal monosomies and the low frequency of certain trisomies, it is thought that these anomalies must lead to very early developmental arrest, most of the zygotes being eliminated before the pregnancy is recognized. This explanation is supported by morphologic examination of zygotes with rare chromosome anomalies; these show very precocious developmental arrests with either complete absence of embryo (blighted ovum) or a malformed embryo of only a few millimeters. This hypothesis of precocious elimination of zygotes with autosomal monosomies and certain autosomal trisomies has recently been experimentally confirmed in translocation of mice (*Mus poschiavinus*) by *Gropp* (1973).

If one accepts the hypothesis that meiotic nondisjunction (in the male or in the female) occurs with equal probability for each chromosome pair, and identical number of zygotes should be conceived with monosomy and trisomy for each chromosome pair and in accordance with the observations of *Gropp* (1973), most of these would be eliminated before pregnancy becomes evident. Thus, the total number of monosomies and trisomies conceived can be estimated from the data for monosomy X and trisomy 16, assuming an equal frequency of nondisjunction for all chromosome pairs. 15% of clinically recognized pregnancies terminate in abortion. (This figure is higher in prospective studies). Thus, for every 1000 clinically recognized pregnancies there are 850 births and 150 abortions. Of these 150 abortions, around 60% (about 100) have a chromosome anomaly. Among these 100 anomalies there are approximately 15 monosomies X and 15 trisomies 16. This leads to an estimate of 15x23 = 345 monosomies and 15x23 = 345 trisomies or a total of 690 anomalies resulting from meiotic nondisjunctions.

To these anomalies can be added the other chromosome anomalies observed in abortions, in particular the triploidies which account for about 20% of the anomalies found in abortions. The resulting figure for total chromosome anomalies conceived is quite close to that of the 850 births at term.

This estimate is in agreement with deductions from other observations in humans: the fundamental work of *Hertig* and *Rock* (1973) on human zygotes of less than 17 days of development and more recently, the studies of *Pearson* et al. (1973) in human spermatozoa.

The figures, though still imprecise, give an idea of the order of magnitude of chromosomal accidents which lead to reproductive failure. This estimate would be around one conception out of every two. While this estimate may be valid for a population as a whole, for each couple the risk of conceiving zygotes with chromosomal anomalies is extremely variable. These variations have been revealed by a study of the obstetric events preceding or following the abortions (*Boué* et al., 1973). This has clearly shown a difference between couples without previous history of abortion (excepting the karyotyped abortion studied) and those with previous abortions. Table 2 summarizes these results.

Table 2. Frequency of recurring abortion after abortion with chromosomal anomalies in relation to previous reproductive history of couples

Previous reproductive history	Frequency of spontaneous abortion in the following pregnancy
No pregnancy	13.3%
Delivery(ies)	7.2%
Delivery(ies) and spontaneous abortion(s)	24.5%
Abortion(s) only and maternal age > 30 yr	41.5%

IV. Phenotypes of Lethal Anomalies

The description of phenotype expressions of lethal chromosomal anomalies is important not only for basic knowledge but for practical reasons since it is unlikely that, due to technical difficulties, cytogenetic analysis of abortuses will ever become a routine procedure. The macroscopic examination of fresh specimens and the histologic study of fixed abortuses may yield information to tentatively diagnose a chromosomal anomaly. In some cases it may be feasible to even specify the type.

Table 3. Mean developmental age of abortuses in relation to type of anomaly

Karyotype	Number	Developmental age (weeks)	Autosomal trisomies	Developmental age (weeks)
Normal	401	5.3	A	3.2
Monosomy	121	5.8	B	4.2
			C	4.3
Trisomy	389	4.2	D	4.7
			E	3.4
Triploidy	135	5.0		
			F	3.4
Tetraploidy	34	2.9		
			G	5.3
			double trisomies	2.9

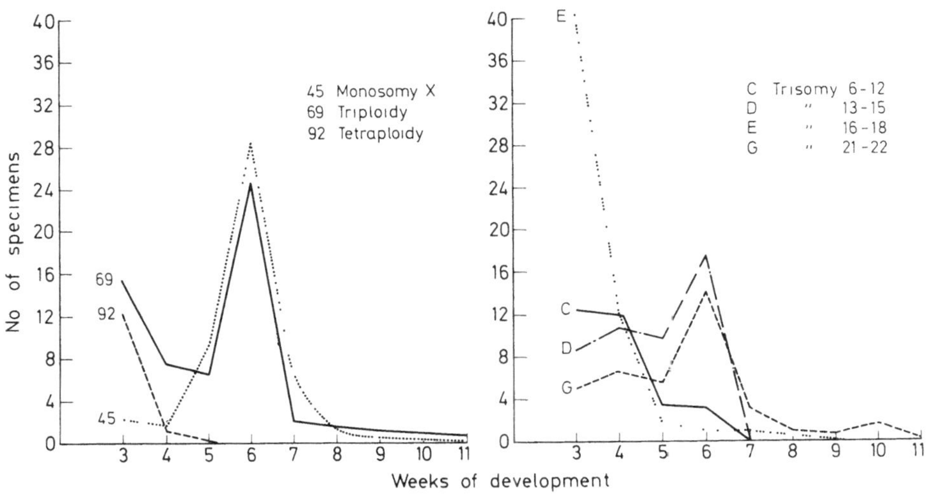

Fig. 3. Developmental age (in weeks) of abortuses with different types of chromosomal anomalies: a. Monosomy X, triploidy and tetraploidy. b. Trisomies C, D, E, and G

In our study phenotype expression is based on three criteria: a) estimation of the developmental age; b) macroscopic and microscopic examinations of the embryo, and c) macroscopic and microscopic examinations of the placenta (*Philippe*, 1974).

Table 3 shows the mean developmental ages of abortuses in relation to the chromosomal anomalies (only specimens of less than 12 weeks were included). Figure 3 represents the number of specimens with the more frequent chromosomal anomalies in each week of developmental age (the curves start only at 3 weeks as very few specimens of a development less than 3 weeks were collected).

In general, always lethal anomalies are characterized by very precocious developmental arrest: tetraploidies, some triploidies, trisomy C, and trisomy 16 (the peak of trisomy E is composed almost exclusively of trisomy 16). Trisomies A, B, and F, rare in abortions, belong to these early developmental arrests. On the other hand, anomalies which are similar to errors compatible with survival to term have a longer duration of development: monosomy X and trisomies D and G.

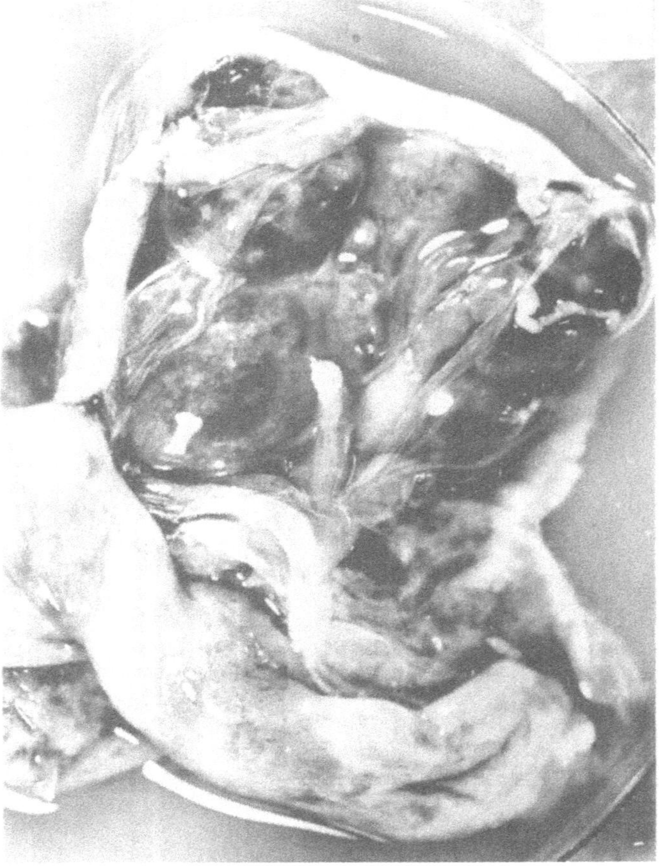

Fig. 4. Specimen with monosomy X. Intact sac (60 mm) of six weeks of development with well-defined cord. At end of cord small amorphous mass of embryonic tissues. Marked subchorionic thrombosis (Breus mole). Gestational age: 92 days

Some chromosomal anomalies may be tentatively diagnosed from phenotypic criteria alone:

In monosomy X developmental arrest occurs at six weeks with few exceptions. Two different phenotype expressions may be observed. The most common, which accounts for two-thirds of the cases, consists of a closed sac with an umbilical cord ending in a small mass with the remnants of macerated embryonic cells (Fig. 4). The yolk sac is present and the chorion is regularly changed to a Breus mole. In the other cases, the appearance of the placenta is the same, but a macerated embryo of a developmental age of 6 weeks is present. It may be morphologically normal, but histological examination frequently reveals a horseshoe kidney. In some of these, it was noted that the developmental arrest of the placenta takes place 1 or 2 weeks before the embryonic developmental arrest. Embryos then are not grossly malformed and death is secondary to placental defects.

Some fetuses with monosomy X have a development of 7-12 weeks and a few are seen in second trimester abortions. They have some common characteristics described by *Singh* (1970): generalized edema, hygroma on each side of the neek, and horseshoe kidney.

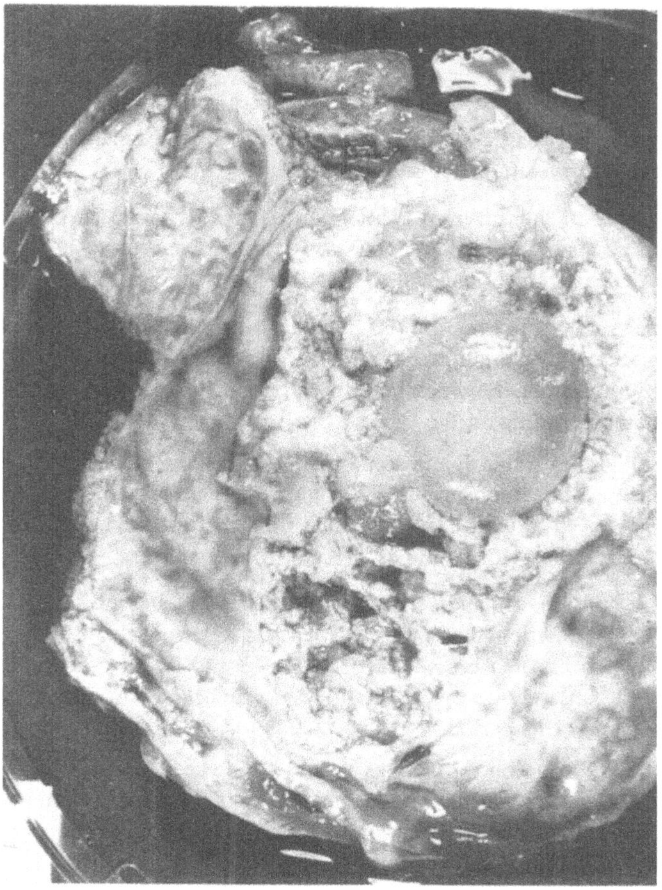

Fig. 5. Specimen with triploidy. Intact sac of 50 mm with disorganized embryo of 4 mm. Marked pseudomolar degeneration of the villi. Gestational age: 184 days

In triploidy the diagnosis is suggested first by the macroscopic examination. It discloses a placenta with cystic chorionic villi associated with an amniotic sac that is either empty or contains an embryo (Fig. 5). There are many intermediate forms, from empty sacs with hydropic pseudomolar villi to stillborn infants with or without gross malformations and associated with a placenta displaying molar degeneration. In our study the mean developmental age was 5 weeks.

Histological examination of the embryo frequently shows malformations of the central nervous system (defects in the division of the brain hemispheres, cyclocephaly, and cebocephaly) or underdeveloped nasal processes and failure of closure of the neural groove (craniorachischisis or spina bifida); absence of hypophysis is frequent.

Histological examination of the placenta reveals voluminous hydropic villi lined with hypoplastic trophoblast which invaginates into the villi to form microcysts.

The main phenotypic feature of triploidy (hydropic molar degeneration with amniotic sac) permits a clear distinction of the triploid conceptuses from the true *hydatidiform mole*. Its characteristics are massive hydatidiform degeneration, lack of embryo or amniotic sac, and trophoblastic hypertrophy at histological examination. In 29 hydatidiform moles, diploid karyotypes were always found, but the sex ratio was 4 with 46,XY to 25 with 46,XX.

In tetraploidy, the development stops between 2 and 3 weeks and the abortus consists of a chorionic vesicle, often without amniotic membrane. Embryonic formations have never been observed in tetraploid specimens (Fig. 6).

Because of the different types of autosomal trisomies and the fact that precise identification of each chromosome pair has been possible only recently, it has often been difficult to collect enough material to describe some characteristics of these trisomies.

The common feature of trisomic placentas is hypoplasia: the villi are rare, frequently avascular, with an underdeveloped trophoblast and few trophoblastic cells. Large cytotrophoblastic cells which have migrated into the stroma are observed in 30-70% of the cases, depending on the type of the trisomy.

In trisomies A and F and some trisomies B and C (C6 and C11) embryonic formations have never been observed. The conceptus is a blighted ovum. One trisomy B was a cyclops of 30-35 days development. In the C group embryonic development of 28-45 days is compatible with trisomies 7, 8, 9, 10, and 12.

Trisomies 13, 14, and 15 usually correspond to an embryo of 40-45 days development. A characteristic phenotype is difficult to describe. Cyclocephaly and underdeveloped nasal process are observed in trisomies 13 and 14 (Fig. 7) and only the appearance of the placenta allows the distinction between these trisomies and triploid specimens aborted at this period.

In the E group only trisomies 16 and 18 are seen, and trisomy 16 represents more than 75% of the specimens. The phenotype of trisomy 16 is characteristic: a chorionic vesicle of 2-3 cm with a small amnion of 5 mm and an embryo of 1 mm or less, arrested at the stage of an embryonic disk.

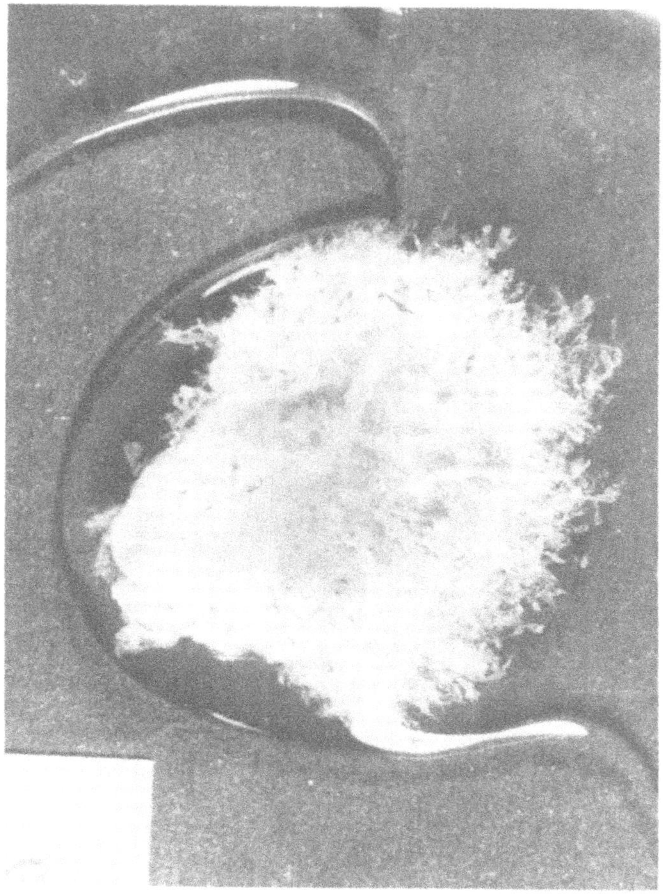

Fig. 6. Specimen with tetraploidy. Empty sac of 20 mm without embryonic formation. Gestational age: 79 days

Embryos with trisomy 21 reach a development of 6-7 weeks; gross abnormalities have never been observed.

The analysis of pathologic lesions observed in abortuses may reveal some general concepts about the mechanisms of abortion. A certain number of anomalies are lethal very early, and one might speculate that the imbalance created by these anomalies in the genetic constitution of these cells is such that embryogenesis, if not completely impossible (as in tetraploidy), is arrested in the early stages (as in trisomies A, B, 16, F . . .). This early arrest of embryogenesis is then the basic phenomenon.

But many anomalies, in particular those which are similar to those observed at birth, are compatible with an embryogenesis which is not sufficiently disturbed to cause the death of the embryo. In these cases a dissociation between the stage of development attained by the embryo and that reached by the placenta is seen. The placental development seems to have stopped more precociously and the death of the embryo is thus secondary.

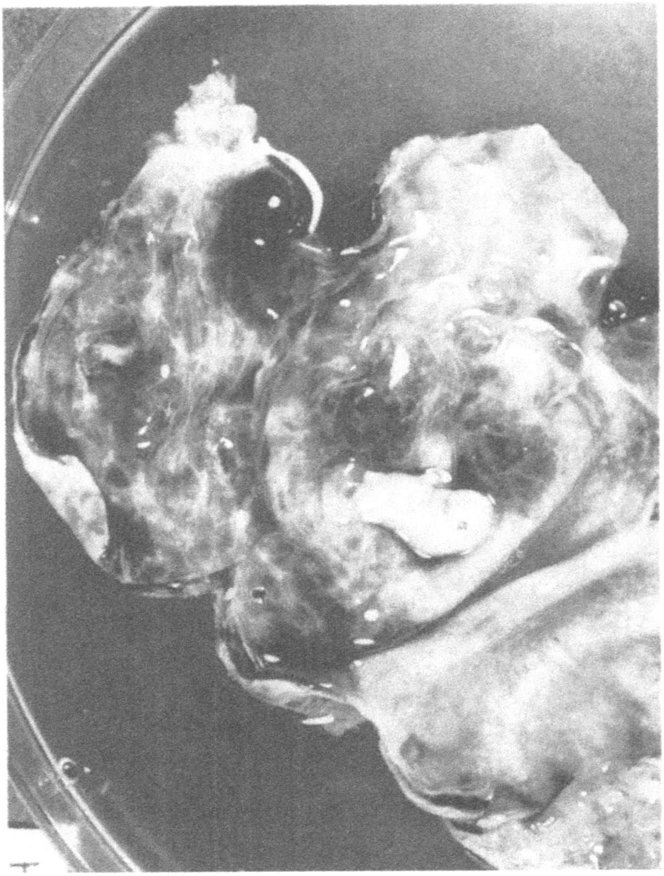

Fig. 7. Specimen with trisomy 14. Intact sac of 60 mm with macerated embryo of 40 days with underdeveloped nasal processes. Hypoplasia of placenta which has developmental age of less than 35 days. Gestational age: 104 days

V. In vitro Growth Characteristics of Aneuploid Embryonic Cells

An analysis of the in vitro growth characteristics of cells from aneuploid embryos has shed some light on the cellular basis for the in vivo arrests of development of these embryos.

Two main approaches were studied: first the longevity and second the doubling time of cultures of aneuploid embryonic cells.

Table 4. Initiation of cell lines from tissues of embryos from spontaneous abortions

Karyotype	Number of primary cultures	Successful primary culture No.	%	1st trypsinization attempted on primary culture No.	%	Initiation of cell line successful (5 or more passages) No.	%
Monosomy X	52	43	83	–[a]	–	(19)[a]	(90)[a]
Trisomy C	28	24	86	16	67	12	75
Trisomy D	51	40	78	27	67	26	96
Trisomy E	43	38	88	25	66	13	52
Trisomy G	42	30	71	19	63	16	84
Triploidy	60	47	78	36	76	32	89
Normal	73	67	92	ND		ND	

ND = Attempts at initiating a cell line were not systematically done for abortuses with normal karyotype.

[a] Initiation of a cell line was only attempted from 21 cultures with monosomy X.

1. In vitro Lifetime

In instances in which a karyologically abnormal zygote was diagnosed from the cells growing from explants of amnion and chorion and in which the primary cultures of embryonic cells were successful, the initiation of a cell line was attempted (Table 4).

Table 5 shows the results of primary cultures and of attempts to initiate cell lines for some of the most frequent chromosomal anomalies. The criterion of success for the initiation of a cell line was the ability to carry the cultures for five or more passages (1:2 split ratios).

Table 5. Initiation of cell lines in relation to the development of the embryo

	Karyotype	Primary culture successful	Primary culture unsuccessful	Attempts of trypsinization	Cell lines initiated <5 passages	Cell lines initiated ⩾5 passages
Embryos ⩽35 days	Trisomy C	14	1	6/14	4	2
	D	5	1	3/5	1	2
	E	26	2	15/26	9	6
	G	4	2	0/4		
	Triploidy	5	2	3/5	1	2
		54	8	27/54	15	12
Embryos >35 days	Trisomy C	10	3	10/10	0	10
	D	35	10	24/35	0	24
	E	12	3	10/12	2	8
	G	26	10	19/26	3	16
	Triploidy	42	11	33/42	3	30
		125	37	96/125	8	88

Successful primary cultures could be initiated from 80% of embryos with abnormal karyotypes. After successful first trypsinization there was a high rate of success in initiating cell lines in monosomy X, trisomies D and G, and triploidy; less frequently in the case of trisomies C and E. From ten embryonic tissues with rare karyotypes (one trisomy A, four trisomies B, two trisomies F and three double trisomies) first trypsinization was performed in six cases. It was successful in four and a cell line was initiated in only two cases (one trisomy F 20 and one double trisomy D+G). In these anomalies with rare karyotypes there is an early arrest in the development of the embryo, usually before the 30th day.

Success in initiating cell lines was then analyzed in relation to the stage of development reached by the embryos from which the cultures were initiated. The rate of success of primary cultures was the same in the two groups: embryos of less and more than 35 days of development. Clear-cut differences appeared in the different steps of initiation of a cell line. In the first group (small embryos) the cell proliferation was sufficient to perform a first trypsinization in only 50% of the cases (27/54) in comparison with 77% (96/125) when the embryo was larger. The culture survived more than 5 passages in only 12 of the 27 first trypsinizations (44%) in the first group and in 88 of 96 (91%) in the second group.

Table 6. In vitro lifetime of aneuploid cells

Karyotype	Number of 1:2 splits before Phase III		
	Min.	Max.	Mean
45 X (n=6)	11	37	20
47 C+ (n=7)	12	18	15
47 D+ (n=11)	18	35	24
47 G+ (n=7)	13	18	15
69 (n=8)	9	40	28

Table 6 shows the lifespan of some of the cell lines, compared with the 40-60 generation "standard" for human diploid cells, nearly all the cell lines with chromosomal anomalies had a reduced lifespan.

The average longevity of the various cell cultures corresponds roughly to the in vivo development attained by the embryo as shown in Figure 3. Cell cultures initiated from those embryos which develop farthest in vivo seem to have the longest in vitro lifespans. There are no in vitro longevity results for those anomalies which have very early in vivo developmental arrests (tetraploidy for example) because in these cases there is usually no embryo, and cell lines cannot be initiated from primary cultures of amnion or chorion. For those anomalies with the most advanced embryonic development cell lines could be initiated from primary cultures 70% of the time (88/125), whereas for those with early developmental arrests, the success rate is only 22% (12/54). It may be that the phenotype effects of certain karyotypes are so drastic that such zygotes can rarely undergo more than a few cell doublings.

2. In vitro Doubling Time

The growth characteristics of 35 embryonic cell lines with chromosome anomalies were compared with seven karyologically normal embryonic cell lines (*Cure* et al., 1974).

The cell lines which were initiated from embryos with normal karyotype from induced or therapeutic abortions (performed for maternal indications) had generation times between 17 and 25 h.

With rare exceptions, embryonic cells with chromosomal anomalies had a longer generation time than normal cells (Table 7). Only 3 of these 35 cell lines had doubling times of less than 30 h.

Table 7. In vitro doubling time of aneuploid cells

Karyotype	Doubling time (h)		
	Min.	Max.	Mean
Normal 46 (n=6)	17	25	21
Abnormal			
45 X (n=6)	33	45	37
47 C+ (n=6)	45	70	50
47 D+ (n=10)	28	44	35
47 G+ (n=5)	30	46	39
69 (n=6)	24	39	32

Thus, cell lines initiated from embryos of spontaneous abortions with chromosomal anomalies exhibit slower generation times and decreased lifespans compared with normal diploid cells. This decreased rate of cell multiplication may be responsible for some of the anomalies of the embryo, particularly those common to several different karyotypes such as underdeveloped nasal processes and certain cardiac abnormalities. A decreased rate of cellular growth and differentiation may lead to asynchronous ontogeny, disturbances in normal embryonic induction processes, as well as to the reduced size and cell numbers in various organs noted by *Naeye* (1967) in infants with autosomal trisomies.

Similarly, *Gropp* in his study on trisomic mouse embryos concludes that the common features are growth retardation, smallness, and hypotrophy of embryo and placenta.

Although we have never been able to initiate cell lines from primary cultures of placental material (amnion and chorion), it has been shown that placentas from spontaneous abortions with chromosomal anomalies are smaller than normal and that this decreased growth is accompanied by decreased placental metabolic activities. This insufficiency of the placenta may influence the growth of the embryo indirectly by producing a nutritional deficiency. Furthermore, hormonal insufficiency of the placenta at the time when hormone production by the corpus luteum is no longer sufficient to maintain the pregnancy may explain why embryos with minor or without malformations that would no interfere with their viability, are aborted at the end of the first trimester (*Boué* et al., 1974).

VI. Summary

Chromosome errors during gametogenesis and at the time of fertilization lead to aneuploid zygotes. It is estimated that about one conception of every two have a chromosome anomaly.

Most of these zygotes have very early arrests of development. The studies of spontaneous abortions show the consequences of some anomalies on early embryogenesis and allow the description of some phenotype expressions.

The analysis of the growth characteristics of cell lines initiated from the tissues of aneuploid embryos shows a decrease in the growth rate of these cells and a reduced lifespan.

References

Boué, A., Boué, J.: Chromosome abnormalities and abortion. In: The Physiology and Genetics of Reproduction. Vol. II (ed. *F. Fuchs*), pp. 317-339. New York: Plenum 1974.

Boué, J., Boué, A., Lazar, P., Guegen, S.: Etude prospective des évènements obstétricaux à la suite d'un avortement caryotypé. In: Chromosomal Errors in Relation to reproductive failure (eds. *A. Boué, C. Thibault*), pp. 391-401. Paris: INSERM 1973.

Boué, J., Boué, A., Spira, A.: Evolution des dosages hormonaux urinaires dans les grossesses avec anomalies chromosomiques létales ou viables. In: Exploration hormonale de la grossesse (ed. *R. Scholler*), pp. 167-189. Paris: S.E.P.E. 1974.

Carr, D.H.: Chromosome anomalies as a cause of spontaneous abortion. Amer. J. Obstet. Gynec. 97, 283-293 (1967).

Cure, S., Boué, J., Boué, A.: Growth characteristics of human embryonic cell lines with chromosomal anomalies. Biomedecine 21, 233-236 (1972).

Gropp, A.: Fetal mortality due to aneuploidy and irregular meiotic segregation in the mouse. In: Chromosomal Errors in Relation to Reproductive Failure (eds. *A. Boué, C. Thibault*), pp. 255-269. Paris: INSERM 1973.

Gropp, A., Putz, B., Zimmermann, U.: Autosomal monosomy and trisomy causing developmental failure. Experimental study (eds.*A. Gropp, E. Grundmann*). Current Topics in Pathology 62, 177-192 (1976).

Hertig, A.T., Rock, J.: Searching for early fertilized human ova. Gynec. Invest. 4, 121-139 (1973).

Jonasson, J., Therkelsen, A.J., Lauritson, J.G., Lindsten, J.: Origin of triploidy in human abortuses. Hereditas 71, 168-171 (1972).

Kajii, T., Ohama, K., Nükawa, N., Ferrier, A., Avirachan, S.: Banding analysis of chromosomal karyotypes in spontaneous abortion. Amer. J. human Genet. 25, 539-547 (1973).

Naeye, R.L.: Prenatal organ and cellular growth with various chromosomal disorders. Biol. Neonate 11, 248-260 (1967).

Pearson, P.L., Geraldts, J.P.M., Pawlowitzki, I.H.: Chromosomal studies on human male gametes. In: Chromosomal Errors in Relation to Reproductive Failure (eds. *A. Boué, C. Thibault*), pp. 219-229. Paris: INSERM 1973.

Philippe, E.: Histopathologie placentaire. Paris: Masson 1974.

Singh, R.P.: Hygroma of the neck in XO abortuses. Amer. J. clin. Path. 53, 104-107 (1970).

Therkelsen, A.J., Grunnet, N., Hjort, T., Myhre Jensen, O., Jonasson, J., Lauritsen, J.G., Lindsten, J., Petersen, B.: Studies on spontaneous abortions. In: Chromosomal Errors in Relation to Reproductive Failure (eds. *A. Boué, C. Thibault*), pp. 81-93. Paris: INSERM 1973.

Uchida, I.A., Holunga, R., Lawler, C.: Identification of triploid genome by fluorescence microscopy. Science 176, 304-305 (1972).

Subject Index

The numbers set in *italics* refer to those pages on which the respective catch-word is discussed in detail.

Index to Volumes 52–61 Current Topics in Pathology

GPSR Compliance
The European Union's (EU) General Product Safety Regulation (GPSR) is a set of rules that requires consumer products to be safe and our obligations to ensure this.

If you have any concerns about our products, you can contact us on

ProductSafety@springernature.com

In case Publisher is established outside the EU, the EU authorized representative is:

Springer Nature Customer Service Center GmbH
Europaplatz 3
69115 Heidelberg, Germany

www.ingramcontent.com/pod-product-compliance
Lightning Source LLC
Chambersburg PA
CBHW061208240926
51846CB00006B/1149
9783642664601